Simulation Foundations, Methods and Applications

Series editor

Louis G. Birta, University of Ottawa, Canada

Advisory Board

Roy E. Crosbie, California State University, Chico, USA
Tony Jakeman, Australian National University, Australia
Axel Lehmann, Universität der Bundeswehr München, Germany
Stewart Robinson, Loughborough University, UK
Andreas Tolk, Old Dominion University, USA
Bernard P. Zeigler, University of Arizona, USA

More information about this series at http://www.springer.com/series/10128

Richard Fujimoto · Conrad Bock
Wei Chen · Ernest Page · Jitesh H. Panchal
Editors

Research Challenges in Modeling and Simulation for Engineering Complex Systems

 Springer

Editors
Richard Fujimoto
Georgia Institute of Technology
Atlanta, GA
USA

Conrad Bock
NIST
Gaithersburg, MD
USA

Wei Chen
Northwestern University
Evanston, IL
USA

Ernest Page
The MITRE Corporation
McLean, VA
USA

Jitesh H. Panchal
Purdue University
West Lafayette, IN
USA

ISSN 2195-2817   ISSN 2195-2825   (electronic)
Simulation Foundations, Methods and Applications
ISBN 978-3-319-86424-2   ISBN 978-3-319-58544-4   (eBook)
DOI 10.1007/978-3-319-58544-4

Preface

A two-day workshop was held on January 13–14, 2016, at the National Science Foundation in Arlington, Virginia, with the goal of defining directions for future research in modeling and simulation and its role in engineering complex systems. The workshop was sponsored by the National Science Foundation (NSF), the National Aeronautics and Space Administration (NASA), the Air Force Office of Scientific Research (AFOSR), and the National Modeling and Simulation Coalition (NMSC) in conjunction with its parent organization the National Training and Simulation Association (NTSA). This book documents the findings emanating from this workshop.

The goal of the workshop was to identify and build consensus around critical research challenges in modeling and simulation related to the design of complex engineered systems—challenges whose solution will significantly impact and accelerate the solution of major problems facing society today. Although modeling and simulation has been an active area of study for some time, new developments such as the need to model systems of unprecedented scale and complexity, the well-documented deluge in data, and revolutionary changes in underlying computing platforms are creating major new opportunities and challenges in the modeling and simulation (M&S) field. The workshop focused on four main technical themes: (1) conceptual models, (2) computational issues, (3) model uncertainty, and (4) reuse of models and simulations.

The workshop resulted in large part from an initiative led by the research and development committee of the National Modeling and Simulation Coalition (NMSC) aimed toward defining a common research agenda for the M&S research community. Recognizing that the modeling and simulation community is fragmented and scattered across many different disciplines, communities, and constituencies, there is a need to gather individuals from different communities to articulate important research problems in M&S. Presentations and panel sessions at several modeling and simulation conferences were held leading up to and following the January workshop to raise awareness of this activity.

Detailed planning began in September 2015 with the formation of the workshop steering committee consisting of Richard Fujimoto (chair, Georgia Tech, and then

NMSC Policy Committee chair), Steven Cornford (NASA Jet Propulsion Laboratory), Christiaan Paredis (National Science Foundation), and Philomena Zimmerman (Office of the Secretary of Defense). An open call was developed and disseminated that requested nominations of individuals, including self-nominations, to participate in the workshop. A total of 102 nominations were received. The steering committee reviewed these nominations, and several rounds of invitations were made until the workshop capacity was reached. Selection of participants focused on goals such as ensuring balance across the four technical theme areas, broad representation from different research communities, inclusion of senior, distinguished researchers in the field, and ensuring inclusion of individuals from underrepresented groups.

A total of 65 individuals attended the workshop. Four working groups were formed, each representing one of the technical theme areas. Participants were initially assigned to one of the working groups; however, attendees were free to participate in a group different from that which the individual was assigned (and some did so), and some chose to participate in multiple groups throughout the course of the two-day workshop. Three individuals within each group agreed to organize and facilitate discussions for that group and help organize the workshop findings.

Each group was charged with identifying the four or five most important research challenges in the specified technical area that, if solved, would have the greatest impact. It was anticipated that within each of these main challenges, there would be some number of key subchallenges that would need to be addressed to attack the research challenge.

Prior to the workshop, several read-ahead documents concerning research challenges in M&S were distributed to the participants. These read-ahead materials included the following:

- National Science Foundation Blue Ribbon Panel, "Simulation-Based Engineering Science," May 2006.
- National Research Council of the National Academies, "Assessing the Reliability of Complex Models, Mathematical and Statistical Foundations of Verification, Validation, and Uncertainty Quantification," 2012.
- A. Tolk, C.D. Combs, R.M. Fujimoto, C.M. Macal, B.L. Nelson, P. Zimmerman, "Do We Need a National Research Agenda for Modeling and Simulation?" *Winter Simulation Conference,* December 2015.
- J.T. Oden, I. Babuska, D. Faghihi, "Predictive Computational Science: Computer Predictions in the Presence of Uncertainty," Encyclopedia of Computational Mechanics, Wiley and Sons, to appear, 2017.
- K. Farrell, J.T. Oden, D. Faghihi, "A Bayesian Framework for Adaptive Selection, Calibration and Validation of Coarse-Grained Models of Atomistic Systems," Journal of Computational Physics, 295 (2015) pp 189–208.
- Air Force Office of Scientific Research and National Science Foundation, "Report of the August 2010 Multi-Agency Workshop on Infosymbiotics/DDDAS: The Power of Dynamic Data Driven Application Systems" August 2010.

In addition, workshop attendees were invited to submit brief position statements of M&S research challenge problems or areas that should be considered for discussion at the workshop. Each proposal was assigned to one of the four technical theme areas and distributed to attendees prior to the meeting.

The workshop program included five application-focused presentations on the first day that described important areas where technical advances in M&S were needed within the context of these domains: sustainable urban growth (John Crittenden), healthcare (Donald Combs), manufacturing (Michael Yukish), aerospace (Steven Jenkins), and defense (Edward Kraft). These presentations, the read-ahead materials, and research challenge proposals submitted by workshop participants were the main inputs used in the workshop.

The remainder of the workshop focused on breakout groups and cross-group discussions with the goal to build consensus around key research challenges that could form the basis for a common research agenda. The first day focused on collecting and consolidating views concerning important research challenges. The second day included brief presentations and discussions reporting progress of the four groups, and further discussion to refine and articulate recommendations concerning research challenges in each of the four technical areas.

This document describes the main findings produced by the workshop. We would like to thank the many individuals and organizations who helped to make this workshop possible. First, we thank the workshop sponsors, especially NSF (Diwakar Gupta) and NASA (John Evans) who provided the principal funding for the workshop. NMSC/NTSA (RADM James Robb) sponsored a reception held at the end of the first day of the workshop, and AFOSR (Frederica Darema) participated in events leading up to the workshop and provided valuable guidance as the workshop was being developed. The five plenary speakers (John Crittenden, Donald Combs, Michael Yukish, Steven Jenkins, and Edward Kraft) provided outstanding, thought-provoking presentations regarding the impact of M&S in their respective application areas. Administrative support for the workshop was provided by Holly Rush and Tracy Scott, and Philip Pecher helped with the development of the final report and Alex Crookshanks help with the graphics used in some of the figures.

Finally, we especially thank the many participants who devoted their time and effort to participate and help develop this workshop report. We thank the group leads for carefully managing the discussions of their groups as well as efforts to organize and, in many cases, write much of the text that is presented here.

Atlanta, GA, USA   Richard Fujimoto
Gaithersburg, MD, USA   Conrad Bock
Evanston, IL, USA   Wei Chen
McLean, VA, USA   Ernest Page
West Lafayette, IN, USA   Jitesh H. Panchal

Although systems arising in the aforementioned applications are very different, they have at least one aspect in common. They are composed of many interacting components, subsystems, and people. Systems such as these that consist of many interacting elements are commonly referred to as *complex systems*. For example, a city can be viewed as a set of infrastructures such as water, energy, transportation, and buildings combined with the social, economic, and decision-making processes that drive its growth and behavior over time. Interactions among the parts of a complex system may give rise to unexpected, emergent phenomena or behaviors that can have desirable consequences, such as the creation of ethnic neighborhoods, or undesirable ones such as urban sprawl. M&S provides critical tools and technologies to understand, predict, and evaluate the behavior of complex systems, as well as the means to develop and evaluate approaches to steer the system toward more desirable states.

Computer-based models and simulations have been in use as long as there have been computers. The value of M&S technologies throughout history is without question. However, the development and use of reliable computer models and simulations is today time-consuming and expensive and can sometimes produce unreliable results. These issues become even more critical as engineered systems increase in complexity and scale and must be deployed in uncertain environments. Advances in M&S technologies *now* are essential to enable the creation of more effective, robust, and less costly engineered complex systems that are critical to modern societies.

Here, we identify key research challenges that must be addressed to enable M&S to remain an effective means to meet the challenges of creating and managing the complex systems that increasingly pervade our society. Key findings of identified challenges fall into four areas:

- *Conceptual modeling.* Understanding and developing complex systems require the collaboration of individuals with widely different expertise. The models through which these individuals communicate and collaborate are commonly referred to as *conceptual models*. Once defined, conceptual models can be converted to computer models and software to represent the system and its behavior. Advances in conceptual modeling are essential to enable effective collaboration and cost-effective, error-free translation of the model into a suitable computer representation.
- *Computational challenges.* Computing and communications technologies have advanced rapidly in the last decade. M&S has not yet fully realized the potential and opportunities afforded by technologies such as mobile and ubiquitous computing, big data, the Internet of Things, cloud computing, and modern supercomputer architectures. This has kept M&S from achieving its fullest potential in modeling complex systems, or being widely deployed in new contexts such as online management of operational systems. Research advances are needed to enable M&S technologies to address issues such as the complexity and scale of the systems that need to be modeled today.

Executive Summary

Engineered systems are achieving unprecedented levels of scale and complexity in the midst of a rapidly changing world with many uncertainties. Cities face enormous challenges resulting from aging infrastructure, increased urbanization, and revolutionary technological changes such as smart electric power grids, photovoltaics and citizen generation of electricity, electrification of the vehicle fleet, autonomous vehicles, and widespread deployment of drones, to mention a few. Forces such as climate change threaten to dramatically impact future developments. The healthcare delivery system faces growing demands for service from an aging population while the system adapts to an explosion in patient medical data, changing payment models, and continued advances in medical technologies. Advances in manufacturing offer the potential for dramatic increases in competitiveness and economic growth, but require rapid increases in automation and fast, seamless integration while new technologies such as additive manufacturing and new approaches to materials design come online. Advanced space missions call for stringent requirements for robustness and flexibility in the face of harsh environments and operation over extreme distances in the presence of environmental surprises, possible technology failures, and heavily constrained budgets. Similarly, defense acquisitions face challenges from asymmetric threats, changing missions, globalization of technology and siloed decision-making processes in the face of declining budgets, a shrinking defense industrial base, and congressional and service imperatives, mandates, and regulations.

In these and many other areas of critical societal importance, modeling and simulation (M&S) plays a key role that is essential to successfully navigate through these challenges and uncertainties. Consideration of alternative futures is inherent in decision making associated with complex socio-technical systems. Empirical investigations of yet-to-exist futures are impossible to realize; however, they can be explored computationally through M&S. Advances in M&S are critical to addressing the many "What if?" questions associated with these and other examples. Advanced modeling techniques, integrated with current and advancing computing power, visualization technologies, and big data sets enable simulations to inform decisions on policies, investments, and operational improvements.

- *Uncertainty.* Models and simulations are necessarily approximate representations of real-world systems. There are always uncertainties inherent in the data used to create the model, as well as the behaviors and processes defined within the model itself. It is critical to understand and manage these uncertainties in any decision-making process involving the use of M&S. New approaches are required to gain better fundamental understandings of uncertainty and to realize practical methods to manage them.
- *Reuse of models and simulations.* It is often the case that models and simulations of subsystems such as the components making up a vehicle are created in isolation and must later be integrated with other models to create a model of the overall system. However, the reuse of existing models and simulations can be costly and time-consuming and can yield uncertain results. Advances are needed to enable cost-effective reuse of models and simulations and to ensure that integrated models produce reliable results.

Findings concerning important research challenges identified in the workshop in each of these areas are discussed in the following and elaborated upon in the subsequent chapters that follow.

A. Conceptual Modeling: Enabling Effective Collaboration to Model Complex Systems

Conceptual modeling is recognized as crucial to formulation and simulation of large and complex problems, but is not well-defined or well-understood, making it an important topic for focused research. Conceptual models are early-stage artifacts that integrate and provide requirements for a variety of more specialized models, where the term "early" applies to every stage of system development. This leads to multiple conceptual models: of reality, problem formulation, analysis, and model synthesis. Developing an engineering discipline of conceptual modeling will require much better understanding of how to make conceptual models and their relationships explicit, the processes of conceptual modeling, as well as architectures and services for building conceptual models.

Finding A.1. Conceptual models must be interpreted the same way by everyone involved, including those building computational tools for these models.

Conceptual models today are most often expressed using some combination of sketches, flowcharts, data, and perhaps pseudo-code. Lack of general agreement on how to interpret these artifacts (i.e., ambiguity) limits the computational assistance that can be provided to engineers. More explicit and formal conceptual modeling languages are needed to support engineering domain integration and analysis tool construction, while retaining accessibility for domain experts via domain-specific modeling languages. Formal conceptual modeling applies not only to the system of interest, but also to the analysis of that system. Several structures have been studied as simulation formalisms; however, there is little consensus on the best approach. Achieving an engineering discipline for M&S will require a more complete set of formalisms spanning from rigorous discrete event, continuous, and stochastic system specification to higher level, perhaps domain-specific, simulation languages.

Finding A.2. Processes for conceptual modeling must meet resource constraints and produce high quality models.

M&S facilities are themselves complex systems, typically requiring multiple steps and decisions to move from problem to solution (*life cycle engineering*). Regardless of complexity, the underlying principle for any type of life cycle engineering is to ensure that unspent resources (e.g., money, time) are commensurate with the remaining work. Reducing uncertainty about work remaining and the rate of resource consumption requires determining the purpose and scope of the system, the kind of system modeling needed (continuous/discrete, deterministic/stochastic, etc.), appropriate modeling formalisms, algorithms, data for calibrating and validating models, and other models for cross-validation. Currently, answering these questions is hampered by a lack of formalized engineering domain knowledge to constrain life cycle decisions and processes. In addition, workflows are central to any approach for making life cycle processes explicit and manageable, but evaluation of these workflows is hampered by the lack of metrics to assess their quality and to assess the quality of the resulting models.

Reducing model defects introduced during the modeling process helps avoid difficult and high-cost amendments of the model as it nears completion. During model development, program leadership must determine what knowledge is to be acquired at each point in the life cycle to maximize value to program stakeholders. Further research is needed on how to set knowledge goals at particular milestones in a system development life cycle. In particular, which knowledge elements are associated with which aspects of the system of interest and its environment? How does one determine the value of acquiring particular kinds of knowledge at particular times in the development life cycle? A complementary approach is to develop a method of measuring the degree of formality and optimization (*maturity*) of M&S processes. No such standardized and systematic assessment methodology is available for M&S processes, but the capability maturity model (CMM) and CMM Integration (CMMI) approach have been applied to many areas, after originating in the software engineering community. Achieving a capability maturity model for M&S processes requires research in a number of areas, including quantitative analysis of the complexity and uncertainties in modeling processes, optimization, risk analysis and control of modeling processes, and quantitative measures of process quality and cost.

Regarding conceptual model validation, the challenge is to find universally applicable concepts, with a theory that is satisfying to all the stakeholders and technology that is germane to a broad set of problems. For example, how does a conceptual model that is suitable for a specific use inform the development of other simulation process artifacts? How do the various stakeholders in the simulation activity use the conceptual model, valid or otherwise? Following the best practice to consider validation early in the development process, advances in theory involving validation of conceptual models will support the rigorous use of conceptual models throughout the simulation life cycle.

Finding A.3. *Architectures and services for conceptual modeling must enable integration of multiple engineering disciplines and development stages.*

Reliable modeling on a large scale for complex systems requires an architecture that enables models to be composed across disciplines. Arriving at such a model architecture requires developing mechanisms for efficient interaction among many sets of laws, determining the level of detail needed to observe emerging behaviors among these laws when integrated, and identifying design patterns appropriate to various communities of interest. The architecture must be supported by services that enable sharing of model elements at all levels, and extension of the architecture as needed. Implementing the architecture and services requires development of integration platforms for modeling, simulation, and execution. One of the major challenges to model integration is the semantic heterogeneity of constituent systems. Simulation integration (co-simulation) has several well-established architectures and standards, but there are many open research issues related to scaling, composition, the large range of required time resolutions, hardware-in-the-loop simulators, and increasing automation in simulation integration. Execution integration is needed as distributed co-simulations are shifting toward cloud-based deployment, a Web-based simulation-as-a-service model, and increased automation in dynamic provisioning of resources.

Reliable model integration depends on sufficient formality in the languages being used. In particular, formalizing mappings between the conceptual models of a system and its analysis models is critical to building reliable bridges between them. Combined with formal conceptual models of both system and analysis, a basis is provided for automating much of analysis model creation through model-to-model transformation. Perhaps the most fundamental challenge in achieving this for conceptual modeling is understanding the trade-offs in recording analysis knowledge in the system model, analysis model, or the mappings between them.

B. Computation: Exploiting Advances in Computing in Modeling Complex Systems

Computing has undergone dramatic advances in recent years. The days are long gone when computers were out of sight of most people, confined to mainframes locked away in machine rooms that could only be operated by highly trained specialists. Today, computers more powerful than yesterday's supercomputers are routinely owned and used by average citizens in the form of smartphones, tablets, laptops, and personal computers. They are key enablers in our everyday lives. Other major technological developments such as big data, cloud computing, the Internet of Things, and novel high-performance computing architectures continue to dramatically change the computing landscape.

Finding B.1. New computing platforms ranging from mobile computers to emerging supercomputer architectures require new modeling and simulation research to maximally exploit their capabilities.

The vast majority of M&S work completed today is performed on traditional computing platforms such as desktop computers or back-end servers. Two major trends in computing concern advances in mobile computing, on the one hand, and the shift to massive parallelism in high-performance computers on the other. As discussed momentarily, exploitation of mobile computing platforms moves models and simulations into new realms where the models interact with the real world. Maximal exploitation of M&S in this new environment, often in conjunction with cloud computing approaches, is not well-understood.

At the same time, modern supercomputer architectures have changed dramatically in the last decade. The so-called power wall has resulted in the performance of single processor computers to stagnate. Improved computer performance over the last decade has arisen from parallel processing, i.e., utilizing many computers concurrently to complete a computation. By analogy, to reduce the time to mow a large lawn, one can utilize many lawn mowers operating concurrently on different sections of the lawn. In much the same way, parallel computers utilize many processors to complete a simulation computation. Modern supercomputers contain hundreds of thousands to millions of processors, resulting in *massively parallel* supercomputers. Further, these architectures are often *heterogeneous*, meaning there are different types of processors included in the machine that have different, specialized capabilities. Effective exploitation of these platforms by M&S programs as well as new, experimental computing approaches is still in its infancy.

Finding B.2. Models and simulations embedded in the real world to monitor and steer systems toward more desirable end states is an emerging area of study with potential for enormous impact.

We are entering an age of "smart systems" that are able to assess their current surroundings and provide useful recommendations to users, or automatically effect changes to improve systems on the fly while the system is operating. For example, smart manufacturing systems can automatically adapt supply chains as circumstances evolve, or smart transportation systems can automatically adapt as congestion develops to reduce traveler delays. Models and simulations driven by online data provide a predictive capability to anticipate system changes and can provide indispensable aids to manage these emerging complex systems. However, key foundational and systems research questions must be addressed to realize this capability. Further, key questions concerning privacy, security, and trust must be addressed to mitigate or avoid unintended, undesirable side effects resulting from the widespread deployment of such systems.

Finding B.3. New means to unify and integrate the increasing "plethora of models" that now exists are needed to effectively model complex systems.

As discussed earlier, complex systems contain many interacting components. Different components often require different types of simulations. For example, some subsystems may be best represented by equation-based, physical system simulations, while others are abstracted to only capture "interesting" events, jumping in time from one event to the next. Simulator platforms, frameworks, tool chains, and standards are needed to allow these simulations to seamlessly interoperate with each other. The simulations may be operating on vastly different time and spatial scales creating mismatches at boundaries where they must interact. Further, many executions of the simulation will usually be required to explore different designs or to assess uncertainties. Many problems call for thousands of runs to be completed. New approaches are needed to complete these runs in a timely fashion.

Finding B.4. Modeling and simulation is synergistic with "big data," and offers the ability to advance predictive capabilities well beyond that which can be accomplished by machine learning technologies alone.

"Big data" analysis techniques such as machine learning algorithms provide powerful predictive capabilities, but are limited because they lack specifications of system behavior. Simulation models provide such specifications, offering the possibility to augment the capability of pure data analysis methods, e.g., to answer "What if?" questions or to be used in non-recurring situations where sufficient data does not exist. There are clear synergies between M&S methods and machine learning algorithms to realize much more effective models that can be used to greatly improve decision making. However, important questions such as effective model and data representation and approaches to create effective integrated models and systems must be addressed to realize this potential.

C. Uncertainty: Understanding and Managing Unknowns in Modeling Complex Systems

All models have inherent uncertainties which limit them from fully explaining past events and predicting future ones. Understanding this uncertainty and its implications is essential in M&S activities.

Finding C.1. There is a need to unify uncertainty-related efforts in M&S under a consistent theoretical and philosophical foundation.

Multiple communities have addressed issues related to uncertainty in models using different mathematical formulations; however, a rigorous theoretical and philosophical foundation is lacking. The lack of such a foundation has resulted in ad hoc approaches for dealing with uncertainty, e.g., use of ad hoc measures of model validity, use of ad hoc models for quantifying model uncertainty, and the artificial distinction between aleatory and epistemic uncertainty. Unification of efforts under a consistent framework is essential for further progress. It is recognized that probability theory is the only theory of uncertainty consistent with the established normative tenets of epistemology. There is agreement that Bayesian probability theory is the consistent foundation for uncertainty in M&S.

Finding C.2. Advancements in theory and methods are needed both for decision making in the M&S process and for M&S to support decision making.

M&S processes are purpose-driven activities which must be considered in the eventual context of use. The specific context defines the role and scope of the model in the decision-making process and the available resources. From this perspective, M&S activities support decision making. Further, modelers are also decision makers who decide how much effort and resources to expend based on the potential impact of the resulting decisions. While decision theory provides the necessary foundation for making M&S decisions, there are unique challenges associated with M&S decisions in an organizational context. Examples of challenges include consistently deducing preferences for individual uncertainty management decisions from overall organizational goals, and the complexity of sequential decisions in M&S.

Finding C.3. Advancements are needed to understand and address aggregation issues in M&S.

M&S of complex systems involves aggregation of information from multiple sources. Techniques such as multi-physics, multi-disciplinary, multi-fidelity, and multi-scale modeling integrate models typically developed by different modelers. Aggregation of information and integration of models is associated with a number of challenges such as seamlessly integrating models across different levels and ensuring consistency in modeling assumptions. Even if consistency across different models is achieved, the fundamental nature of aggregation can also result in erroneous results due to the path dependency problem. There is a need to address the challenges associated with aggregation of physics-related and preference-related information in modeling complex systems.

Finding C.4. While there has been significant progress on understanding humans as decision makers, the utilization of this knowledge in M&S activities has been limited.

Humans are integral elements of socio-technical systems. Accurately modeling human behavior is essential to simulating overall system behavior. Further, the developers and users of models are human decision makers. Therefore, the effectiveness of the model development and usage process is highly dependent on the behavior of the human decision makers. Better understanding of biases that exist in human decisions can help toward better designed control strategies for socio-technical systems, better M&S processes, more efficient allocation of organizational resources, and better model-driven decision making. Addressing human aspects in M&S would require collaboration between domain-specific modeling researchers and researchers in social, behavioral, and psychological sciences.

Another key challenge in M&S is communication of model predictions and associated uncertainty among stakeholders. There is a need for techniques for consistently communicating the underlying assumptions and modelers' beliefs along with their potential impact on the predicted quantities of interest. There is a need for bringing uncertainty at the core of educational curricula. A modern curriculum on probability in engineering and science is needed to equip students with the foundation to reason about uncertainty. Finally, while "big data" has been used to inform the models of the simulated system, the use of "big data" has introduced new challenges associated with incomplete or noisy samples, high dimensionality, "overfitting," and the difficulties in characterizing uncertainty in extrapolative settings and rare events. New research approaches that incorporate rigorous mathematical, statistical, scientific, and engineering principles are therefore needed.

D. Reuse of Models and Simulations: Reducing the Cost of Modeling Complex Systems

As discussed earlier, the ability to reuse models and simulations can substantially reduce the cost of creating new models. This topic overlaps with some of the challenges discussed earlier. For example, the challenges of conceptual modeling and computational challenges play pivotal roles in reuse processes as well. However, the facets highlighted in those areas focused on identifying reusable solutions, selecting the best reusable solution under the given conceptual and technical constraints, and the integration of the identified solution into the appropriate solution framework, all while taking organizational and social aspects into account. The following three challenges were identified to categorize solution contributions: (1) the theory of reuse, (2) the practice of reuse, and (3) social, behavioral, and cultural aspects of reuse.

Finding D.1 Advancements in the theory of reuse are needed to provide a firm theoretical foundation for producing robust and reliable reuse practices.

A firm theoretical foundation is needed for producing robust and reliable reuse practices. While heuristics and best practices can guide practitioners successfully, it is the theory behind these approaches that ensures their applicability. While good heuristics and practices have worked well elsewhere and have led to good results, only a theoretical framework can provide formal proofs of general validity. In support of these tasks, key questions are related to composability, the use of metadata to enable reuse, and opportunities for reuse automation.

Finding D.2 Guides of good practices on reuse of simulation solutions, data, and knowledge discovery can in particular support the workforce.

Although work in recent years contributed to solving various challenges to the day-to-day practice of reuse in modeling, good practices are still needed to support the simulation workforce. To this end, important topics include the reuse of M&S, the reuse of data, and the reuse of knowledge management. M&S research addresses primarily issues confronting the reuse of representations of models and their implementation in simulation languages and frameworks. Research on data reuse focuses on input needs and output possibilities of simulation systems, as well as the necessary metadata approaches. Finally, knowledge management research is coping with general challenges applicable to all these topics.

Finding D.3 Research on social, behavioral, and cultural aspects of reuse shows that they may stimulate or impede reuse at least as much as technical constraints.

Several recent studies show that often intangible human and organizational factors hinder the reuse of models, simulations, and data, even when all conceptual and technical aspects can be solved. Key research questions have the objective to identify and teach the skills necessary for a model or simulation producer to increase the ease of reuse by others if the producer chooses to, and can afford to do so. Programmatic issues, questions on risk and liability, and general social and behavioral aspects must be better understood and disseminated to contribute to reuse practices.

In summary, solutions to the research challenges described across the four technical areas discussed above will greatly expand our abilities to design and manage complex engineered systems. Advances in M&S will have broad impacts across many of the most important and challenging problems facing society today. The world is rapidly becoming more and more interconnected and interdependent, resulting in consequences that are increasingly more difficult to anticipate, impeding planning and preparation. While M&S has served us well in the past and is a critical tool widely used today, new advances are essential for the technology to keep pace with a rapidly changing world and create new capabilities never even considered in the past.

Contents

Contents

Chapter 1
Introduction

Richard Fujimoto and Margaret Loper

Computer-based models and simulations are vital technologies needed in advanced economies to guide the design of complex systems. M&S technologies are essential to address the critical challenges facing society today such as the creation of smart, sustainable cities, development of advanced aircraft and manufacturing systems, and creating more secure and resilient societies and effective health care systems, to mention a few.

However, the development and use of reliable computer models and simulations is today time consuming and expensive, and results produced by the models may not be sufficiently reliable for their intended purpose. M&S faces unprecedented new challenges. Engineered systems are continually increasing in complexity and scale. Advances in M&S are essential to keep up with this growing complexity and to maximize the effectiveness of new and emerging computational technologies to engineer the increasingly complex systems that are needed in the future.

1.1 Why Now?

Computer-based models and simulations have been in use as long as there have been computers. For example, one application of the ENIAC, the first electronic digital computer, was to compute trajectories of artillery shells to create firing tables used in World War II. There is no question that computer simulations have had major impacts on society in the past and will continue to do so in the future.

R. Fujimoto (✉)
Georgia Institute of Technology, Atlanta, USA
e-mail: fujimoto@cc.gatech.edu

M. Loper
Georgia Tech Research Institute, Atlanta, USA
e-mail: Margaret.Loper@gtri.gatech.edu

© Springer International Publishing AG (outside the USA) 2017   1
R. Fujimoto et al. (eds.), *Research Challenges in Modeling
and Simulation for Engineering Complex Systems*, Simulation Foundations,
Methods and Applications, DOI 10.1007/978-3-319-58544-4_1

The importance of M&S was recognized in the U.S. House of Representatives that declared it a critical technology of national importance (U.S. House of Representatives 2007).

New developments in M&S technologies are of critical importance now. M&S applications are rapidly increasing in scale and complexity as systems become more complex and interconnected. For example, consider the use of M&S to inform policymakers to steer urban growth toward more sustainable trajectories. It is widely recognized that one must view cities as a whole and consider interdependencies among critical infrastructures such as transportation, water, and energy, as well as interactions with social processes and policy. Each of these systems and infrastructures is a large, complex adaptive system in its own right. Creating simulation models able to capture the behaviors and interactions among these infrastructures and social-economic processes is even more challenging. Advances in modeling and simulation technologies are needed.

While the emerging demands of new applications built using M&S present one set of challenges, the underlying computing platforms and technologies exploited by M&S have undergone dramatic changes in the last decade, also highlighting the need for technological innovations. These advances create new opportunities, and challenges for modeling and simulation to achieve even greater levels of impact. Trends such as the Internet of Things and "Big Data" have strong implications concerning the future of modeling and simulation. Online decision-making is an area of increasing importance with the emergence of mobile computing and growth in technologies such as sensor networks. Modeling and simulation are complementary to the exploitation of machine learning. While models derived purely from machine learning algorithms offer much benefit, they do not include behavioral descriptions of the system under investigation that are necessary for prediction of dynamic system behaviors and enable what-if experimentation, or analysis of situations where sufficient data, or the right data are not available, e.g., due to privacy or other concerns. At the same time, power and energy consumption has become an important consideration in computing, both for mobile computing platforms and computing in data centers. Massively, parallel multiprocessor systems containing over a million cores, GPUs, and cloud computing have emerged in importance in the last decade, motivating research to effectively exploit these platforms. Cloud computing offers much broader exploitation of M&S technologies by making high-performance computing capabilities much more broadly accessible, and embedding simulations into operational environments presents new opportunities and challenges.

1.2   Modeling and Simulation

There are several definitions of models, simulations, and the M&S discipline. The U.S. Department of Defense (DoD) defines these terms as follows in their online glossary (MSCO 2016):

- *Model*: a physical, mathematical, or otherwise logical representation of a system, entity, phenomenon, or process.
- *Simulation*: a method for implementing a model and behaviors in executable software.
- *Modeling and simulation* (M&S): the discipline that comprises the development and/or use of models and simulations.

Here, we are specifically concerned with computer models where the representation is stored and manipulated on an electronic computer. A simulation captures salient aspects of the dynamic behavior of the modeled system over time. Typically, a simulation model captures the state of the system being modeled at one instant in time through a set of values assigned to variables and data structures in the computer program, commonly referred to as state variables. For example, a simulation of a transportation system might define state variables for each vehicle in the system indicating its current location, direction of travel, speed, acceleration, etc. A set of procedures or programs transform these state variables to represent the state of the system from one time instant to the next. In this way, the simulation constructs a trajectory or sample path of the state of the system over the period of time that is of interest.

We note that modeling and simulation are closely related, but distinct areas. *Modeling* is primarily concerned with the representation of the system under investigation. Models always involve a simplified representation. Therefore, a key question concerns what is included, and by implication, what is left out of the model. *Simulation* is concerned with transforming the model to mimic the behavior of the system over time. Key questions include the algorithms, procedures, and software that are required to perform this transformation. In some cases, creation of the model is of primary concern, and simulation may be secondary or not required at all. For example, when creating the design of an automobile to be handed over to a factory responsible for manufacturing, the dynamics of the vehicle as it is travelling on the roadway are not important. Here, we are concerned with both modeling and simulation aspects of complex engineered systems.

The modeling and simulation discipline covers many aspects. The elements that are most relevant to the discussion here are perhaps best described within the context of the life cycle of an M&S project or study. The process depicted in Fig. 1.1 captures the basic elements of this life cycle (Loper 2015). The life cycle begins by defining the purpose and scope of the study. Specific questions concerning the actual or envisioned system under investigation are defined. The purpose and scope forms the basis of the *conceptual model* that characterizes the system under investigation. Elaborated upon below, the conceptual model includes descriptions of the abstractions used to describe the system; key assumptions used by the model are defined explicitly or (more often) implicitly in the conceptual model as well as key inputs and outputs. Data used to characterize the system and information concerning important processes are collected, analyzed, and incorporated into the model. The conceptual model is then converted into a simulation model and computer program. Verification is concerned with ensuring that the simulation program is an acceptable representation of the conceptual model. Verification is to a large extent a software development activity. Validation is concerned with ensuring the simulation program is an acceptable representation of the system under investigation for the questions posed in the study. This is often accomplished by comparing results produced by the simulation program with data measured from the system under investigation or, in the absence of an implemented system that can be measured, other models of the system. Once the simulation model has been validated to an acceptable degree of certainty, it is applied to answer the questions posed in the first step of the life cycle. The model will be executed many times, e.g., using different random number streams for stochastic simulation models, or to explore various parameter settings; the experiment design defines the simulation runs that are to be completed. Output analysis concerns characterization and quantification of model results, e.g., to determine confidence intervals and variance of output values. Simulation models often must be modified and evolve during the life cycle, e.g., to improve the validity of its results or to incorporate new capabilities or to answer new questions not recognized in the initial design. Configuration control refers to the processes necessary to manage these changes. Finally, once the necessary results have been produced, they must be documented and presented to the individuals or decision makers to illustrate key behaviors and outcomes predicted by the simulation model.

The boxes in blue in Fig. 1.1 represent model development activities and the orange boxes represent simulation development activities. These boxes do not represent an absolute separation between modeling and simulation—the "develop simulation model and program" box bridges between the modeling and simulation

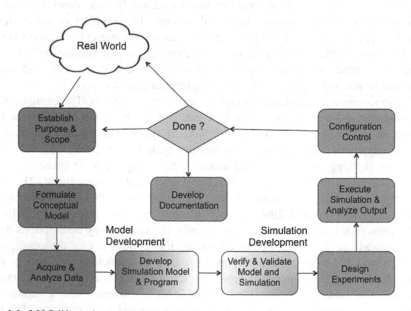

Fig. 1.1 M&S life cycle process (Loper 2015)

activities, and the "verify and validate model and simulation" box represents activities that are performed throughout the entire life cycle.

Here, we focus on four key aspects of the life cycle, discussed next:

- Development of the conceptual model.
- Computational issues concerning the execution of simulation models.
- Understanding and managing uncertainty inherent in models.
- Reuse of models and simulations to accelerate the simulation model development process.

1.2.1 Conceptual Model

A model is a simplification and approximation of reality, and the art of modeling involves choosing which essential factors must be included, and which factors may be ignored or safely excluded from the model. This is accomplished through the process of simplification and abstraction. Simplification is an analytical technique in which unimportant details are removed in an effort to define simpler relationships. Abstraction is an analytical technique that establishes the essential features of a real system and represents them in a different form. The resultant model should demonstrate the qualities and behaviors of a real-world system that impact the questions that the modeler is trying to answer. The process of simplification and abstraction is part of developing the conceptual model. A simulation conceptual model is a living document that grows from an informal description to a formal description and serves to communicate between the diverse groups participating in the model's development. It describes what is to be represented, the assumptions limiting those representations, and other capabilities (e.g., data) needed to satisfy the user's requirements. An informal conceptual model may be written using natural language and contain assumptions about what is or is not represented in the model. A formal conceptual model is an unambiguous description of model structure. It should consist of mathematical and logical relationships describing the components and the structure of the system. It is used as an aid to detect omissions and inconsistencies and to resolve ambiguities inherent in informal models, and is used by software developers to develop code for the computational model.

1.2.2 Simulation Development and Reuse

Once the conceptual model has been created, the next step is to create the simulation by coding the conceptual model into a computer recognizable form that can calculate the impact of uncertain inputs on decisions and outcomes that are important relative to the purpose and scope of the study. Translating the model into computer code and then into an executable involves selecting the most appropriate simulation methodology and an appropriate computer implementation. Methodologies appropriate for modeling complex systems include the following: discrete event simulation, discrete event system specification (DEVS), petri nets, agent-based modeling and simulation, system dynamics, surrogate models, artificial neural networks, Bayesian belief networks, Markov models, game theory, network models (graph theory), and enterprise architecture frameworks, among others.

The development of new simulation programs can be greatly accelerated by reusing existing simulations rather than developing everything "from scratch" for each new model. At its most basic level, common components of the program such as key data structures and libraries for random number generation can be readily reused rather than redeveloped. A much more ambitious goal is to reuse entire model components or entire simulations. Many large, complex systems may be viewed as collections of subsystems that interact with each other in some fashion. A simulation model of such "systems-of-systems" may be derived by integrating existing simulation models of the subsystems. The end goal is to create simulations that may be easily composed with other simulations, much like composing mathematical functions.

1.2.3 Simulation Model Execution

A single execution of the simulation model is commonly referred to as a trial. A simulation study will typically require many hundreds or thousands of trials. Each trial is an experiment, for instance, where we supply numerical values for input variables, evaluate the model to compute outcomes of interest, and collect these values for later analysis. Exhaustive exploration of all input parameters, e.g., to identify an optimal solution is usually impractical due to the large number of runs that would be required. Further, for stochastic models where random numbers are used to characterize uncertain variables, the output of a simulation run produces a single sample. Hence, simulation often relies on random sampling of values for the uncertain variables. To obtain more accurate results the number of trials may be increased, so there is a trade-off between accuracy of the results and the time taken to run the simulation.

The platform on which the simulation executes is an important consideration in executing the simulation model. For large models, parallel processing techniques utilizing computing platforms composed of many processors may be used to accelerate model execution. In other contexts, the simulation may be used to control an operational system. In this case, data from the system is collected and fed directly into the simulation model. The simulation may then analyze alternate options and produce recommended courses of action that are then deployed in the operational system. This feedback loop may be automated or may include human decision makers. This paradigm of utilizing online data to drive simulation computations and to use these results to optimize the system or adapt the measurement process is referred to as dynamic data-driven application systems (DDDAS).

1.2.4 Uncertainty and Risk

A simulation model is always an approximate representation of reality. As such, there are always uncertainties concerning the relationship between the model and the actual system. Uncertainty can enter mathematical models and experimental measurements in various contexts. For example, parameter uncertainty comes from the model parameters that are inputs to the mathematical model, but whose exact values are unknown and cannot be measured or controlled in physical experiments, or whose values cannot be exactly inferred by statistical methods. Parametric variability comes from the variability of input variables of the model. Structural uncertainty, aka model inadequacy, model bias, or model discrepancy, comes from the lack of knowledge of the underlying true physics. Algorithmic uncertainty, aka numerical uncertainty, comes from numerical errors and numerical approximations per implementation of the computer model. Interpolation uncertainty comes from a lack of available data collected from computer model simulations and/or experimental measurements. For other input settings that do not have simulation data or experimental measurements, one must interpolate or extrapolate in order to predict the corresponding responses.

A quantitative risk model calculates the impact of the uncertain parameters and the decisions that are made on outcomes that are of interest. Such a model can help decision makers understand the impact of uncertainty and the consequences of different decisions. The process of risk analysis includes identifying and quantifying uncertainties, estimating their impact on outcomes that are of interest, building a risk analysis model that expresses these elements in quantitative form, exploring the model through simulation, and making risk management decisions that can help us avoid, mitigate, or otherwise deal with risk.

1.3 Preliminary Questions

Five topic areas are noted to generate challenges for modeling and simulation research:

1. Selected applications that would benefit from advances in modeling and simulation
2. Conceptual modeling
3. Computational methods: algorithms for simulation and other types of inference
4. Uncertainty in modeling and simulation
5. Reuse of models and simulations

   Important questions in each of these areas are discussed next.

1.3.1 Applications

Engineered systems continue to grow in complexity and scale. Existing modeling and simulation capabilities have not kept pace with the need to design and manage new emerging systems. Although the focus here is on modeling and simulation per se, distinct from the domain in which the technology is applied, the requirements of modeling and simulation technologies are ultimately derived from the application. In this context, the new emerging developments in specific applications of societal importance are relevant in order to assess the needs and impacts that advances in modeling and simulation will have within those domains.

Specific application domains discussed later include the following:

- Aerospace
- Health care and medicine
- Manufacturing
- Security and defense
- Sustainability, urban growth, and infrastructures

1.3.2 Conceptual Modeling

Although one of the first steps in the development of a model is the development of its conceptual model, such conceptual models have traditionally been informal, document-based. As the complexity of simulation models increases and the number of domain experts contributing to a single model grows, there is an increasing need to create formal, descriptive models of the system under investigation and its environment. This is particularly important for the engineering of complex systems where multiple system alternatives are explored, compared, and gradually refined over time. The descriptive model of each system alternative—describing the system of interest, the environment, and interactions between them—can serve as a conceptual model for a corresponding analysis or simulation model. Formal modeling of these descriptive, conceptual models poses significant research challenges:

- How can models expressed by different experts in different modeling languages be combined in a consistent fashion?
- What level of formality is suitable for efficient and effective communication?
- What characteristics should a modeling environment have to support conceptual modeling in an organizational context—a distributed cognitive system?
- What transformations of conceptual models to other representations are possible and useful? What are the major impediments to realizing such transformations?

1.3.3 Computational Methods

The main reason for modeling is to extend human cognition. By expressing our knowledge in a mathematical formalism, the rules of mathematical inference implemented in computer algorithms can be used to draw systematic conclusion that are well beyond the natural cognitive ability of humans. For instance, simulation allows us to project how the state of a system will change over time for complex systems with millions of state variables and relationships. Advancing the algorithms for such inference so that ever larger models can be processed more quickly is likely to remain a crucial capability for engineering and science. Besides simulation, there is an increasing role for model checking, especially for engineered systems that are affected by high-impact low-probability events.

This raises questions such as the following:

- What are current trends in computing affecting modeling and simulation and how can they best be exploited?
- How will these trends change the nature of simulation and reasoning algorithms?
- What are the major gaps in computational methods for modeling and simulation, and what are the most important research problems?
- How can one best exploit the vast amounts of data now becoming available to synergistically advance M&S for engineering complex systems?

1.3.4 Model Uncertainty

The goal of modeling and simulation often is to make predictions, either to support decisions in an engineering, business, policymaking context or to gain understanding and test hypotheses in a scientific context. It is impossible to prove a model is correct—the predictions are always uncertain. Yet, many models and simulations have been proven to be useful, and their results are routinely used for many purposes. To further improve the usefulness of models, it is important that we develop a rigorous theoretical foundation for characterizing the uncertainty of the predictions. Within the modeling and simulation community, there is still a lack of agreement on how best to characterize this uncertainty. A variety of frameworks have been proposed around concepts of validation and verification, and a variety of uncertainty representations have been proposed.

This leads to the following questions:

- What is the most appropriate approach to consistently represent and reason about uncertainty in complex systems?
- What is the best approach to characterizing the uncertainty associated with a simulation model in order to enable and facilitate reuse?

- How should one aggregate knowledge, expertise, and beliefs of multiple experts across different domains?
- What is the best approach to take advantage of the large and diverse datasets for characterizing uncertainty and for improving model accuracy?
- What are the most promising approaches to accelerate the validation of models for specific application contexts?

1.3.5 Model Reuse

Although modeling has become indispensable in engineering and science, the cost of creating a good model can be considerable. This raises the question of how these costs can be reduced. One approach is to encode domain knowledge into modular, reusable libraries of models that can then be specialized and composed into larger models. Such a modular approach allows the cost of model development, testing, and verification to be amortized over many (re)uses. However, reuse also introduces new challenges:

- How can a model user be confident that a planned reuse of the model is within the range of uses intended by the model creator?
- How can one characterize the uncertainty of a model that is reused (possibly with some adaptations to a new context)?
- How can one characterize the uncertainty of simulation models obtained through the composition of multiple models?
- How can one accelerate the process of adapting and reusing models for different purposes? What are the fundamental limitations of technologies for model reuse?

The chapters that follow discuss each of these topics. Chapter 2 reports on the five aforementioned application areas. Chapter 3 reports on discussions and research challenges concerning conceptual models. Chapter 4 describes computational challenges in M&S. Chapter 5 discusses uncertainty and associated research challenges. Finally, Chapter 6 characterizes and presents research challenges concerning the reuse of models and simulations.

Long established as a critical technology in the design and evaluation of systems, modeling and simulation is now at a critical crossroads. M&S technologies face increasing challenges resulting from the scale and complexity of the modern engineered systems that need to be designed, understood, and evaluated. At the same time, technological advances offer the potential for M&S technologies to not only meet these challenges, but also to provide even greater value in new areas than is offered today. It is clear that through the Internet, social networks, and other advances, the world is rapidly becoming more interconnected and complex, and the pace of change is accelerating. While M&S has served society well in the past, new innovations and advances are now required to enable it to continue to be an indispensable tool to enable deep understandings and effective design of new and emerging complex engineered systems.

References

Loper, M. (editor). 2015. Modeling & simulation in the systems engineering life cycle: Core concepts and accompanying lectures, Series: Simulation Foundations, Methods and Applications. Springer; 14 May.

Modeling and Simulation Coordination Office (MSCO). 2016. DoD M&S Catalog. Accessed April 21, 2016. http://mscatalog.msco.mil/.

U.S. House of Representatives. House Resolution 487. July 16, 2007.

Chapter 2
Applications

William Rouse and Philomena Zimmerman

2.1 Introduction

Modeling and simulation provide a powerful means to understand problems, gain insights into key trade-offs, and inform decisions at all echelons of the domain. Applications of modeling and simulation should be driven by the nature of the problems of interest and the appropriateness of the model or simulation for the problem and domain in which this approach is being considered or applied.

This chapter begins by reviewing five important areas to understand the nature of the problems addressed rather than the approaches to modeling and simulation employed in these instances. This leads to consideration of crosscutting challenges associated with these examples. This chapter concludes with a discussion of specific modeling and simulation challenges identified.

2.2 Five Examples

Five exemplar application areas where modeling and simulation can provide the means to understand problems, gain insights into key trade-offs, and inform decisions include the following:

W. Rouse (✉)
Stevens Institute of Technology, 1 Castle Point Terrace,
Hoboken, NJ 07030, USA
e-mail: rouse@stevens.edu

P. Zimmerman
Office of the Deputy Assistant Secretary of Defense for Systems Engineering,
Washington DC, USA
e-mail: philomena.m.zimmerman.civ@mail.mil

© Springer International Publishing AG (outside the USA) 2017
R. Fujimoto et al. (eds.), *Research Challenges in Modeling
and Simulation for Engineering Complex Systems*, Simulation Foundations,
Methods and Applications, DOI 10.1007/978-3-319-58544-4_2

- Urban infrastructure
- Health care delivery
- Automated vehicle manufacturing
- Deep space missions
- Acquisitions enterprise

Table 2.1 compares these five examples in terms of the nature of the problem addressed rather than the specific modeling and simulation employed. The five examples are contrasted in terms of top-down forces, bottom-up forces, human phenomena, and the difficulty of the problem.

2.2.1 Top-Down Forces

The top-down forces affecting urban infrastructure include the consequences of climate change, forced migration, and macroeconomic trends. In contrast, health care delivery is being affected by increased demand for services from an aging population, increased prevalence of chronic disease, and changing payment models. Many of these forces are exogenous to the urban and health care enterprises.

Automated vehicle manufacturing is being affected by demands from the Department of Defense for rapid design, development, manufacturing, deployment, and sustainment. This occurs in the broader context of the acquisitions enterprise, which is being affected by congressional and military services' imperatives, mandates, regulations, and budget pressures. These forces are endogenous to the defense enterprise, but exogenous to particular programs.

The top-down forces affecting deep space missions include mission requirements for robustness and flexibility, as well as the magnitude and timing of budgets. These requirements are seen as exogenous to the extent that they are taken as nonnegotiable. There could be, of course, trade-offs between requirements and budgets.

2.2.2 Bottom-Up Forces

Bottom-up forces tend to come from within the enterprise and hence can be seen as endogenous to the system. Such forces are often more amenable to prediction, control, and perhaps design. Thus, they are more likely to be explicitly represented in models and simulations rather than seen as being external to the phenomena being modeled.

The bottom-up forces of increased demands on infrastructure and generation of waste, as well as dealing with waste, affect urban infrastructure. Health care delivery must deal with patients' disease incidence, progression, and preferences, as well as providers' investment decisions. Deep space missions are affected by environmental surprises, technological failures, and public support for space

Table 2.1 Comparison of five applications examples

	Urban infrastructure	Health care delivery	Automated vehicle manufacturing	Deep space missions	Acquisitions enterprise
Top-down forces	Consequences of climate change, forced migration, and macroeconomic trends	Increased demand for services, increased prevalence of chronic disease, and changing payment models	Demands for rapid design, development, manufacturing, deployment, and sustainment	Mission requirements for robustness and flexibility, magnitude, and timing of budgets	Congressional and services' imperatives, mandates, and regulations; budget pressures
Bottom-up forces	Increased demands on infrastructure and generation of waste, dealing with waste	Patients' disease incidence, progression, and preferences; providers investment decisions	State of technology for design, development, and manufacturing; availability of tools, components, and materials	Environmental surprises, technological failures, and public support for space exploration	Asymmetric threats, changing missions, globalization of technology, and declining defense industrial base
Human phenomena	Social and political forces, individual preferences, and decisions regarding consumption and use of infrastructure	Disease dynamics, patient choice, and clinician decisions	Design and development decision making, supervisory control of manufacturing, and operation and maintenance of deployed systems	Design and development decision making, ground operations decision making	Decision making at all levels; sustainment of deployed systems
Difficulty of problem	Fragmented decision making across city, state, and federal agencies; aging infrastructure	Uncertainty of demands for various services, science and technology advances, and stability of payment models	Required pace of rapid automation exceeds state of the art, level of integration of all needed ingredients very demanding	Harsh environment, extreme distances, communications delays of minutes to hours, and infeasibility of maintenance and repair	Plethora of models, methods, and tools; fragmented and siloed decision making

exploration. These three examples concern the magnitudes and uncertainties associated with demands on those systems.

Automated vehicle manufacturing is affected by the state of technology for design, development, and manufacturing, as well as the availability of tools, components, and materials. Acquisitions enterprises must address asymmetric threats, changing missions, globalization of technology and, in some areas, the declining defense industrial base. These two examples are laced with changing requirements and both technological and organizational constraints.

2.2.3 Human Phenomena

Behavioral and social phenomena are much more difficult to model than purely physical systems. The five examples differ significantly in terms of the prevalence of human phenomena.

Social and political forces, as well as individual preferences and decisions regarding consumption and use of infrastructure affect urban infrastructure. Disease dynamics, patient choice, and clinician decisions affect health care delivery. Many of the behavioral and social phenomena associated with these examples are not amenable to design changes.

Automated vehicle manufacturing is laced with design and development decision making, supervisory control of manufacturing, and operation and maintenance of deployed systems. Deep space missions are similarly affected by design and development decision making, as well as ground operations decision making. Acquisitions enterprises are also affected by decision making at all levels, and sustainment of deployed systems. The decision making for these three examples is often amenable to various levels of decision support.

2.2.4 Difficulty of Problem

The difficulty of addressing urban infrastructure is exacerbated by fragmented decision making across city, state, and federal agencies, all in the context of severely aging infrastructure. Health care delivery is difficult due to uncertainty of demands for various services, impacts of science and technology advances, and stability of payment models. These two examples face uncertain demands and organizational difficulties.

For automated vehicle manufacturing, the required pace of rapid automation exceeds the state of the art. The level of integration of all needed ingredients is very demanding. Acquisitions enterprises are beset by a plethora of models, methods, and tools, as well as fragmented and siloed decision making. Deep space missions face harsh environments, extreme distances, communications delays of minutes to hours, and infeasibility of maintenance and repair. These three examples are laced with technological and technical difficulties.

2.2.5 Crosscutting Issues

In all cases, the problem being addressed must be considered within the broader enterprise context of top-down and bottom-up forces that influence the problem and likely constrain the range and nature of solutions, as well as the choice of the model (s) or simulation(s) to be applied. In other words, what phenomena are internal and external to the model and simulation?

Many models and simulations do not incorporate rich representations of the human behavioral and social phenomena associated with the problems of interest. Yet, human operators and maintainers, as well as citizens and consumers, are central to several of the example problems. Humans provide flexible, adaptive information processing capabilities to systems, but also can make risky slips and mistakes. There is much more uncertainty in systems where behavioral and social phenomena are prevalent.

There are also the human users of models and simulations, ranging from direct model-based decision support to use of model-derived evidence to support organizational decision processes. Technology now enables powerful decision support environments that can empower decision makers to immerse themselves in the complexity of their problem spaces. Evidence of this is increasingly immersive interactive visualizations that prompt expressions like "wow," but are not well understood in terms of their impacts on decision making.

All of the examples are plagued, to a greater or lesser extent, by the fragmentation and incompatibilities of the ever-evolving range of available tools. Some areas such as computational fluid dynamics, semiconductor design, and supply chain management have achieved a level of standardization, but this is quite difficult in areas where "one off" solutions are the norm. Investing in developing and refining a model and simulation is easier to justify when one is going to produce thousands or even millions of the system of interest. This is more difficult to justify and accomplish well when the target is, for example, a single mission.

Underlying all five examples are implicit assumptions and questions about the model or simulation of interest. Is the credibility of a model or simulation understood, accepted, or implied? Are the effects of uncertainty understood? Can one trust in the results of the model or simulation? Can truly emergent behavior be elicited by the representation(s) chosen? How can one understand the current configuration as the model or simulation evolves? Does the model or simulation conform to exchange standards that enable valid conjunctions of models or simulations?

There is also an assumed demand for interactivity between the users and the model or simulation environment. This is likely to require more intelligence and resilience in the model or simulation to enable valid responses to the range of external stimuli allowed. At the very least, it requires that developers of models and simulations have deep understanding of the use cases the model is intended to support as well as the likely knowledge and skills of the envisioned users.

Finally, a major challenge concerns the necessary regulatory, statutory and cultural hurdles that must be surmounted to actually use a model or simulation, and of the set of phenomena associated with the problem of interest to support making real decisions. This requires that decision makers both trust the model or simulation and be willing to make the decisions being informed by the visualizations of model outputs for the scenarios explored.

2.3 Modeling and Simulation Challenges

The applications cited above are part of an almost infinite space of uses for models and simulations. There are overlaps in the application of models and simulations; overlaps in the necessary characteristics of the model or simulation for the intended use; overlaps in the methods and processes used to develop models or simulations; and overlaps in the challenges with the application of modeling or simulation.

The development of a model, or a simulation execution of a model, as a representation of reality can only go so far. Most problems are complex, and hence are decomposed to enable a solution. Modularization of a problem so that each part can be modeled or simulated is fairly straightforward. What is not straightforward is the understanding of the interdependencies between the system modules being modeled. In part, this is caused by the loss of understanding of these interdependencies when a system is decomposed, or modularized. You cannot validly model or simulate what you do not understand.

Because of the loss of understanding of important interdependencies, it is very difficult to explicitly and adequately represent the interactions in the models of the decomposed system. Because of this, it is not possible to recompose the models or simulation executions into a representation of the original system. Emergent behaviors as a result of the composition may or may not replicate the unidentified relationships between the modules of the original system. In other words, the emergent behaviors may be artifacts of the decomposition rather than reflections of reality.

The challenges with emergent behavior extend beyond the composition of models or simulation executions. These challenges extend into the relationships which exist between the modeled physical and organizational phenomena, and the simulation of the processes in which the models are to be used. This boundary point can be thought of simply as an interface definition.

The concept of an interface is simple; however, the necessary depth of information needed to express the relationship between the physical and organizational phenomena and the system or process that uses them is not easily identified. Methods for identifying the needed depth of information, based on understanding of the interactions, are an area of significant challenge in efficient use of conceptual modeling or execution of conceptual models in a simulation.

Continuing with issues associated with the interactions between modeled parts of a system, or models within a larger system, challenges exist with automated methods for constructing an operational environment from a hierarchical set of model components, for example, within a product line. Considering the needed depth of information, there are challenges in knowing how much information to include in the operational environment. This multifaceted problem includes identification of the necessary depth of information to properly exercise the model, or gain the necessary data from the simulation execution. There are no known methods for translating between the system and the environment in which it operates. As stated earlier, you cannot validly model what you do not understand.

Other challenges in conceptual modeling exist in translating the descriptive models from their representational format into executable simulations. These challenges exist in both the essence of model content and the computer environment in which the model will execute. For example, some conceptual models exist in text format. The automated translation of a model expressed in a rich language, into an environment which ultimately is expressed in Boolean expressions, is perhaps the largest of the challenges in the translation domain. Less complex, but no less challenging, is the ability to completely describe the model or simulation so that automated methods can, without loss, translate from one representational format to another.

Additional challenges in modeling and simulation exist within the computational environment in which they exist. Just as there are challenges in modeling the relationships between modeled parts of a system, there are dependencies which exist between the model or simulation, and the infrastructure in which it operates or exists. This is especially true for simulations. An improper execution environment will introduce unquantified unknowns into the results. Potentially less known is the impacts to the model from the infrastructure in which it exists. The model, as the basic representation of reality, is assumed to be uncorrupted. The model is usually never assessed for representational accuracy or corruption effects when accessed. It is assumed to be in the same state as when it was last 'touched'. The ability to assure that the model is free from infrastructure-induced defects is a gap existing today.

Once the model is put into use, challenges exist due to the need to match the results to the user's viewpoint. Visualization of the model or visualization of the simulation results can be assessed as correct or incorrect simply because the visualization tools used do not represent the results in a manner that is understandable, or useful to the user. Work remains to be done on characterizing the user needs and preferences, as matching that to the visualization effects of the model, or data set resulting from the simulation execution. These challenges can be extended deeper than the visualization tool. Characterizing the user needs and matching them to the models or simulation execution that fits the problem space in an automated fashion has the potential to significantly increase the efficiency in the use of the model or simulation.

Other challenges exist in the representational format of the underlying phenomena within the model or simulation. There exists a plethora of representational methods for models. Not all of these are known to all model builders or users. Model or simulation users need to be able to assess the applicability of models or simulations to various problems, which exist in formats that are unknown or less known to the users. Methods to translate model or simulation characteristics from one format to another, or represent them in a standard, acceptable format, remain a challenge today.

Challenges existing at the intersection of model and simulation content and the infrastructure in which it exists or executes include the need for methods to identify optimal fidelity or resolution needed for proper application to decision support. Typically, decision makers express their needs in terms of textual or spoken questions. This hides the complexity which exists in matching computational simplicity and rigor to the rich context underlying written or spoken format. Beginning with simple noun–verb comparisons will get us part of the way to the match. However, nouns and verbs are not easily matched to mathematical expressions which exist in the computational environment. Methods are needed both to automatically perform the match and to break down the language question into constituent parts which more easily match the computational component, taking care to allow for variability in the language itself. Early steps include allowing the user to base model or simulation selections on the presentation of computational expressions.

Challenges remain in the representation of the natural environment, both internally and externally. Biological and social processes are not easily expressed using logic constructs. As such, a different tactic may be to express what is known in logical constructs, and to quantify what is not known. This serves to reduce the problem to some extent but leaves unresolved a way to quantify the uncertainty of biological and social systems.

Particularly, challenging is a method to express environments that are driven by human behavior, such as socioeconomic environments. There is a lack of methods to express, or understand, what is not expressed in systems and environments where humans are involved. Human actions can be unscripted, unpredictable, and often not possible to model in ways comparable to physical phenomena. This is partly because of the unknown relationships, but also because human judgement can be quite subtle.

In order to model or simulate interactions involving biological (human, animal, etc.) inputs, or human–human interactions results, the multi-fidelity, multimodal, multi-domain models often constructed involve rather mixed precision. The ability to actually do this, and have a repeatable, predictable result is necessary, but methods do not exist today to accomplish this, or validate the composition or decomposition.

The challenges discussed, thus, far have not included the challenges coming from the application domains themselves. One challenge is with the applicability of the model or simulation beyond the problem space for which it was originally intended. Models and simulations are often reused due to word of mouth, with or without the associated documentation. Challenges exist with models or simulations, built for one purpose being validly used in another domain. Just because it was not built for a particular purpose does not mean that it is inherently not usable for another purpose. The challenge is how to validate a model in a different domain.

Modeling and simulation exist for almost every activity today. However, each activity domain retains its own language. This usually underlies the domain's models and simulations. Challenges exist with integrating the domain language and knowledge, extended into the model or simulation manifested. Integration or interaction between multiple domains is usually accomplished using language. This allows for reasoning and translation of concepts. How can this be extended to facilitate multi-domain model integration?

Many models and simulations are never retired. Such models and simulations evolve through modification. Is it possible to characterize the types of modifications performed to evolve the models? If yes, how? When is it necessary to characterize a model as new? When is it impossible to assume validation due to changes? There exists a need to answer these questions, since evolving models and simulations need to be trusted.

A final challenge remains in understanding and then describing a model or simulation as a complete entity, for future use, for contracting purposes, etc. The methods to completely describe a model or simulation begin with understanding what "complete" means in a domain, as well as use of a model or simulation. The use of a model or simulation within a domain, or ecosystem, needs to be articulated to fully understand the boundary conditions of the model or simulation, the extensibility of the model or simulation, the history of the model or simulation, and the current state of use.

Chapter 3
Conceptual Modeling

Conrad Bock, Fatma Dandashi, Sanford Friedenthal,
Nathalie Harrison, Steven Jenkins, Leon McGinnis,
Janos Sztipanovits, Adelinde Uhrmacher, Eric Weisel and Lin Zhang

3.1 Introduction

Over the past decade in the modeling and simulation community, there has been a growing interest in and concern about "conceptual modeling." Generally accepted as crucial for any modeling and simulation project addressing a large and complex

C. Bock (✉)
National Institute of Standards and Technology, Gaithersburg, USA
e-mail: conrad.bock@nist.gov

F. Dandashi
The MITRE Corporation, Mclean, USA
e-mail: dandashi@mitre.org

S. Friedenthal
SAF Consulting, Reston, USA
e-mail: safriedenthal@gmail.com

N. Harrison
Lawrence Livermore National Laboratory, Livermore, USA
e-mail: Nathalie.Harrison@drdc-rddc.gc.ca

S. Jenkins
NASA Jet Propulsion Laboratory, Pasadena, USA
e-mail: sjenkins@jpl.nasa.gov

L. McGinnis
Georgia Institute of Technology, Atlanta, USA
e-mail: leon.mcginnis@isye.gatech.edu

J. Sztipanovits
Vanderbilt University, Nashville, USA
e-mail: janos.sztipanovits@vanderbilt.edu

A. Uhrmacher
University of Rostock, Rostock, USA
e-mail: lin@informatik.uni-rostock.de

© Springer International Publishing AG (outside the USA) 2017   23
R. Fujimoto et al. (eds.), *Research Challenges in Modeling
and Simulation for Engineering Complex Systems*, Simulation Foundations,
Methods and Applications, DOI 10.1007/978-3-319-58544-4_3

problem, conceptual modeling is not well defined, nor is there a consensus on best practices. "Important" and "not well understood" would seem to qualify conceptual modeling as a target for focused research.

One may define conceptual models as "early stage" artifacts that integrate and provide requirements for a variety of more specialized models. In this view, conceptual models provide a foundation from which more formal and more detailed abstractions can be developed and eventually elaborated into analysis models (e.g., for simulation). However, "early" and "late" are relative terms that apply within each stage of development. For example, creating an analysis model might involve describing (i.e., modeling) the analysis independently of software ("conceptually") before implementation and execution. As a consequence, there might be multiple "early" models: conceptual models of reality and conceptual models of analysis; and there may be multiple versions of conceptual models as the understanding of the target system matures and the analysis design and implementation evolves.

These varieties of conceptual models are sometimes distinguished in existing work, with different terminology. In 2013, Robinson used "conceptual model" to mean "a non-software specific description of the simulation model, … describing the objectives, inputs, outputs, content, assumptions, and simplifications of the [simulation] model" and "system description" to mean models derived from the "real world," with two stages of computer-specific models derived from the system description (Robinson 2013). In a 2012 tutorial, Harrison and Waite use "conceptual model" to mean "an abstract and simplified representation of a referent (reality)" (Harrison and Waite 2012), instead of Robinson's "system description."

With this context, developing an engineering discipline of conceptual modeling will require much better understanding of:

1. how to make conceptual models explicit and unambiguous, for both the target system (or referent) and the target analysis,
2. the processes of conceptual modeling, including communication and decision-making involving multiple stakeholders,
3. architectures and services for building conceptual models.

Answering the first question (explicitness) requires considering alternative formalisms for expressing conceptual models, and the languages based on these formalisms, which are addressed in Sect. 3.1. The second question (process) is discussed in Sect. 3.2. The third question involves architectures for model engineering, as well as services provided to conceptual modelers, and is covered in Sect. 3.3.

E. Weisel
Old Dominion University, Norfolk, USA
e-mail: eweisel@odu.edu

L. Zhang
Beihang University, Beijing, China
e-mail: johnlin9999@163.com

3.2 Conceptual Modeling Language/Formalism

An articulated conceptual model, whether describing the system of interest (the *referent*, in Robinson's terminology) or an analysis model of the system of interest, is expressed using some language, which may be formal or informal, graphical, textual, mathematical, or logical. Today, the situation is that most often, conceptual models are expressed using some combination of sketches, flowcharts, data, and perhaps pseudo-code. Lack of general agreement on the implications of these techniques (i.e., ambiguity) limits the computational assistance that can be provided to engineers. Incorporating conceptual modeling into a modeling and simulation engineering discipline will require more explicit and formal conceptual modeling languages. However, conceptual modeling must be done in a manner accessible to domain engineers, who might not be trained in the necessary formalisms. This is addressed in the first subsection below. In addition, formal conceptual modeling applies as much to analysis as to the referent systems, raising questions about the variety of approaches to simulation, as covered in the second subsection. Formality in model integration is discussed in Sect. 3.3.

3.2.1 Domain-Specific Formalisms

In mathematical logic, formalism is the application of model and proof theory to languages, to increase confidence in inferring new statements from existing ones (Bock et al. 2006). In practice, however, most mathematicians are more informal in their definitions and proofs, with peer review confirming results, or not. We expect conceptual modeling formalisms to be rigorous approaches to studying referent and analysis models, at least in the sense of mathematical practice. Formal approaches have fewer, more abstract categories and terms than less formal ones, facilitating integration across engineering domains and construction of analysis tools. However, by using more abstract language, formal approaches are often too far from the common language of applications to be easily understood by domain experts and too cumbersome to use in engineering practice, e.g., in air traffic control, battlespace management, healthcare systems, and logistics. More specific formalisms would be useful not only to domain experts, for describing their systems, but also to technical or modeling experts who must translate the system description into analysis models and maintain them, and to other stakeholders who may need to participate in validation.

Logical modeling is a widely used approach to formalizing domain knowledge (often called *ontology*, more specifically description logics Baader et al. 2010). Ontologies can support acquisition of increasing levels of detail in model structure and also education and communication. For example, in modeling an ecosystem, one begins with words and phrases expressed in natural language, such as pond, organism, bio-matter, and insect. Some words will represent categories or classes, while others represent instances falling into those categories. Also, words that connote action will reflect behaviors that are at the core of dynamic system specification. Words and phrases can be connected through relationships forming semantic networks and concept maps. Semantic networks (see, e.g., Reichgelt 1991) grew out of theories of cognition around associative memory (Quillian 1968), whereas concept maps (see Novak 1990) grew out of a theory of associative networks for the purpose of learning, both essential for capturing expert knowledge. Both are closely related to description logics (Sattler 2010; Eskridge and Hoffman 2012).

Developing explicit and formal conceptual models of the referent will require ontologies and a suitable knowledge representation. Contemporary modeling languages have been proposed and used for modeling software systems (UML OMG 2015b), for general systems modeling (OPM Reinhartz-Berger and Dori 2005; SysML OMG 2015a), and for modeling systems in the military domain (UPDM/UAF (OMG 2016), DoDAF (US Department of Defense 2016a), MoDAF (U.K. Ministry of Defense 2016)). Domain-specific modeling languages (DSMLs) also have been developed, for modeling business processes (OMG 2013), for modeling biological systems, SBML (Finney et al. 2006), SBGN (Le Novère et al. 2009), and others. Of course, general purpose languages can be specialized to a domain as well. In addition to using ontologies for domains, methods, and processes, DSMLs for representing knowledge along with induced constraints and interdependencies will also help to reduce uncertainty in the modeling process, e.g., to answer questions like what modeling approach, execution algorithm, or steady-state analyzer to use. Thus, ontologies and DSMLs for modeling and simulation methods are relevant definitely in the requirements stage, where it is decided which formalism to use and how to execute a model, and also for validating and verifying a large set of methods. Suitable ontologies, if in place, will help in identifying solutions.

While these developments are an important element of establishing an engineering discipline of modeling and simulation, they do not yet go far enough. Ontology is not sufficiently applied to formal and domain-specific modeling languages, leaving a major gap in linking formalisms to engineering domains. Many of the available models of formal and domain languages only categorize terminology rather than semantics of the terms, and consequently cannot utilize domain knowledge to increase the efficiency of formal computations or bring results from those computations back into the domains. For example, ontologies are available for Petri Nets (PN), a widely used simulation formalism, but these only formalize terminology for reliable interchange of PN models, rather than enabling a uniform execution of them across tools (Gaševića and Devedžić 2006). In addition, if adequate DSMLs do not already exist within a domain, each application modeler still must develop a problem-specific ontology and capture problem-specific knowledge in a DSML. Within a particular domain, e.g., logistics, the creation of a domain-specific ontology and modeling framework would support all modelers within that domain (Huang et al. 2008); (McGinnis and Ustun 2009); (Thiers and McGinnis 2011); (Batarseh and McGinnis 2012); (Sprock and McGinnis 2014).

Work on modeling formal and domain-specific languages, including semantics as well as terminology and how to integrate them for practical use, are in its early stages (Mannadiar and Vangheluwe 2010); (Bock and Odell 2011), but several results have emerged during the past decade. This is an important area for future research and development.

Language modeling (*metamodeling*) has become a widely used method for precisely defining the abstract syntax of DSMLs (the part of syntax that omits detailed visual aspects of a language). A metamodel is the model of a modeling language (Karsai et al. 2004), expressed by means of a metamodeling language (Flatscher 2002). There are several metamodeling languages in practical use today, ranging from informal, graphical languages, such as UML class diagrams and OCL used by the Object Management Group (OMG 2015b) (OMG 2014), the Eclipse Modeling Framework (Eclipse Foundation 2016a), or MetaGME (Emerson and Neema 2006). A formal metamodeling language based on algebraic datatypes and first order logic with fixpoint is FORMULA from Microsoft Research (Jackson and Sztipanovits 2009).

Metamodeling can be used to specify diagrammatic syntax of DSMLs, in conjunction with their abstract syntaxes above. For example, languages such as Eclipse's Graphical Modeling Project (Eclipse Foundation 2016b) and WebGME (Institute for Software Integrated Systems 2016) provide a graphical metamodeling environment, as well as auto-configuration into domain-specific modeling environments using the metamodels created there. Metamodeling of diagrammatic syntax also enables standardized interchange between tools and rendering of graphics (Bock and Elaasar 2016).

Metamodeling has a role in precisely defining the semantics of DSML's. For example, FORMULA's constraint logic programming capability is used for defining semantics of DSMLs via specifying model transformations to formal languages (Simko et al. 2013). The portion of UML's metamodel that overlaps description logic can be extended to specify patterns of using temporal relation models in UML, providing a basis to formalize the semantics of UML's behavioral syntaxes (Bock and Odell, 2011).

Other approaches to the development of DSMLs include developing a DSML for a subset of the Simulink language, by defining the operational semantics, rather than by creating a meta-model (Bouissou and Chapoutot 2012), or in a similar vein, developing a DSML for systems biology based on an abstract syntax and operational semantics (Warnke et al. 2015).

3.2.2 A Unified Theory for Simulation Formalisms

Conceptual modeling applies not only to the system of interest, but also to the analysis of that system. Our understanding of a system of interest evolves from our earliest concept of it as we gain deeper understanding through the development of system models. In the same way, our understanding of the analysis itself also may evolve as we better understand the system of interest and begin to elaborate our analysis model. To support conceptual modeling of simulation analysis, it seems reasonable that we should first have the ontology, semantics, and syntax to formally define a simulation. Unlike the case of other analyses, such as optimization, this requirement has not yet been satisfied for simulation. Several structures have been studied as simulation formalisms; however, there is little consensus on the best approach. In the same way that various models of computation provide a basis for theory within computer science, considering various simulation formalisms will further the development of a robust theory of simulation.

Some formalisms are available for general discrete event simulation, some adopted industrially and others not. For example, the DEVS language (Zeigler et al. 2000) provides a mathematically precise definition of discrete event systems, and there also are a number of computational implementations, so it is unique in providing both a simulation programming language and an associated mathematical specification. It is not widely used industrially, however, in part, perhaps, because of the requirement to express all behavior using state machines. Popular discrete event simulation languages or environments, such as Arena (https://www.arenasimulation.com), FlexSim (https://www.flexsim.com), Simio (http://www.simio.com), and Tecnomatix Plant Simulation (https://goo.gl/XmQGgN), provide a programming language with semantics and syntax, but not a corresponding formal definition. In part, this is due to the intent of many commercial simulation languages to support simulation in a particular domain, such as Tecnomatix Plant Simulation, which is naturally reflected in the semantics of the languages.

Another line of research is to view simulations through the lens of dynamical systems and computational complexity theory. This is particularly suitable when studying complex socially coupled systems. Formal computational and mathematical theory based on network science and graphical dynamical systems has been studied in Mortveit and Reidys (2008), Barrett et al. (2004, 2006), Adiga et al. (2016), Rosenkrantz et al. (2015). The theoretical framework allows one to study formal questions related to simulations, including: (i) computational lower and upper bounds on computing phase space properties, (ii) design questions: how does one design simulations to achieve a certain property, (iii) inference questions: how does one understand the conditions that led to the observed behavior.

Achieving an engineering discipline for modeling and simulation will require a more complete set of formalisms spanning up from rigorous discrete event, continuous, and stochastic system specification to higher level, perhaps domain-specific, simulation languages. In some areas, those domain-specific modeling languages that combine a rigorous mathematical semantics with a convenient modeling tool are already in use, e.g., in the area of cell biology, or collective adaptive systems (often based on a continuous time Markov chain semantics). For example, some specialized simulation languages for biology are based on mathematical formalisms, such as ML-Rules (Helms et al. 2014), Kappa (Harvard Medical School 2016), or BioNetGen (BioNetGen 2016), among others. In general, however, this still represents a very significant challenge for the modeling and simulation community.

3.3 Conceptual Model Development Processes

Model development is a challenging and highly intricate process, with many questions needing to be answered, as discussed in this section. Currently, answering these questions in a systematic and informed manner is hampered by a lack of formalized knowledge in the modeling domains and in modeling and simulation in general. Providing these would constrain development decisions and the design of development processes themselves, reducing uncertainty in model life cycle engineering. The first subsection below gives background on model development processes and analyzes questions about them. The next two subsections (effectiveness and maturity) describe complementary approaches to reducing model defects introduced during the modeling process. These help avoid difficult and high-cost amendments of the model after it is finished. It is impossible to reduce model defects to zero during development, leading to the need for validation after the model is built, the results of which are also useful during model development, as addressed in the last subsection. Taken together, progress in these areas can significantly enhance the credibility of models by improving the quality of processes that produce them.

3.3.1 Motivation and Research Approach

The purpose of modeling and simulation is to improve our understanding of the behavior of systems: An executable model M of a system S together with an experiment E allows the experiment E to be applied to the model M to answer questions about S (Cellier 1991). Simulation is fundamentally an experiment on a model. A conceptual model C is the articulated description of S, upon which both M and E are developed. In science we seek to understand the behavior of natural systems; in engineering we seek to design systems that exhibit desired behavior. Because modeling and simulation facilities are themselves complex systems, it is seldom possible to go in one step from problem to solution. The processes involved in modeling and simulation require different degrees of human interaction, different computer resources, are based on heterogeneous, partly uncertain knowledge defined more or less formally, and involve different types of expertize and users. Data, knowledge, processes, and orchestration vary depending on the system to be modeled, the questions to be answered, and the users. In these processes different versions of models and artifacts are generated, that need to be put into relation to each other.

Model life cycle Engineering (MLE) captures the highly iterative process of developing, verifying, validating, applying, and maintaining a model. MLE is an area that requires significant study and exploration to meet society's needs and problems. How is MLE different than Engineering Design or Software Engineering life cycles? In some instances, it may be possible to build on these related engineering fields in our attempt to forge MLE as a subdiscipline of Modeling and Simulation (M&S) . It is expected that MLE will contain phases for constructing models and simulations by beginning with requirements and then proceeding to other phases such as design, analysis, implementation, verification and validation (V&V), and maintenance.

MLE concepts and methods should not be limited to developing M and E; they also should be applied to the conceptual model C, describing S and used in developing both M and E. Clearly, this requires that C be expressed in a form that enables MLE concepts and methods to be applied.

The underlying principle for any type of life cycle engineering, however, is to ensure that unspent resources (e.g., money, time) are commensurate with work remaining. For complex systems with substantial de novo content, there is typically considerable uncertainty in both the work remaining and the rate of resource consumption. Resources are therefore held in reserve to protect against depletion due to undesired outcomes. Bearing these principles in mind, a life cycle approach for model/simulation development should include answering the following questions:

- Purpose and Scope Characterization: Who are the stakeholders of the model? What are their concerns? In particular, what are the specific aspects of system behavior we seek to understand through modeling? The answers form the context for the relevant conceptual models. Identification of stakeholders and concerns is a complex undertaking involving a broad spectrum of disciplines, including perhaps the political and behavioral sciences. For example, a macroeconomic simulation of energy production, distribution, and consumption would rightly recognize the public at large as a stakeholder, but it would be counterproductive to ask individual citizens simply to enumerate their concerns since ordinary citizens are not likely to understand the stake they have in atmospheric carbon dioxide or sulfur dioxide. Consequently, it may be necessary to develop methods that combine opinion research with education and outreach to designated proxies for the public interest.
- Phenomena Characterization: Is the referent a continuous system, discrete event system, or discrete stepwise system? Are stochastic or spatial aspects important? What are the elements of the system which contribute to the behavior of interest? What scientific disciplines address the behavior of interest? Answers to these kinds of questions will help to identify the content of the conceptual model and perhaps how it should be expressed. Having identified concerns, it is not necessarily simple to determine the scope of scientific phenomena to adequately address those concerns. For example, if stakeholders are concerned about the availability of drinking water, it may under some circumstances suffice to consider only hydrological phenomena. Under other circumstances it may be necessary to consider also social, economic, and political phenomena. Decisions will ultimately of course involve judgment, but research may elucidate principles and techniques that might prove useful for such analysis.

- Formalism Characterization: What formulations will be most appropriate to describe the relevant system elements and characterize the phenomena of interest in the form of input-output relations? The conceptual model must support these formulations. The choice of formalism will depend on the nature of the system being modeled, as determined by phenomena characteristics above. Once the nature of the system is identified, how is it best described, e.g., for a continuous system, are block diagrams most appropriate, or systems dynamics, or an object-oriented approach like Modelica (Modelica Association 2014b)? What mathematical formulations will be used to characterize the phenomena of interest in the form of input-output relations? Differential equations? Statistical models? Logical models? A given phenomenon may be mathematically characterized in different ways, depending upon, among other things, the nature of the concerns under consideration. If we are primarily concerned with long-term average behavior, we might choose a lumped-parameter description that assumes all short-term variation self-cancels over time. On the other hand, if we are concerned with infrequent extreme events, we will require a characterization that captures higher-order dynamics accurately. Research may help us better understand how to infer the possible mathematical formalism from a given referent model, but also how to develop requirements for the referent model from a useful mathematical formalism.
- Algorithm Characterization: What solution algorithms will be selected for computing the input-output relations? What verification test cases are appropriate? Since the conceptual model is a bridge from S to the computational model M, it may be important to understand and accommodate the specific target algorithmic implementation. For example, canonical linear least squares problems can be stated more compactly as so-called normal equations, but in practice, normal equations are more computationally complex to solve than the better-behaved and more efficient orthogonal triangularization. These are considerations of numerical analysis, a mature field with decades of sound theory and practical technology, but needing integration into M&S methodologies.
- Model Calibration: What data is available to calibrate and later validate the model, M? Is it necessary to calibrate a conceptual model, and if so, how is it done? How does one validate a conceptual model?
- Cross-validation: Do other models exist with which the new model can be cross-validated? If there are other existing conceptual models, how can they be compared to support cross-validation?

These questions need to be addressed during the requirements phase of the model engineering life cycle. However, answers are likely to be revised during the subsequent phases. From this point on, conventional software development life cycle considerations apply. In addition, special consideration needs to be given to validation and verification of model variants and their interdependencies. Research is needed to understand how to help answer the above questions: how to manage the evolution process of a model and the data, knowledge, activities, processes and organizations/people involved in the full life cycle of a model?

Managing the life cycle process of a model is one of the most important tasks of model engineering. Some research topics should be attacked, for example, how to structurally describe the modeling process, and how to identify the characteristics of activities involved in model construction and management to ensure improvement of model quality and development efficiency and reduction of full model life cycle cost.

Some decisions, e.g., which execution algorithm to select, might even be supported automatically by exploiting machine learning methods (Helms et al. 2015). However, automatic solutions to these decisions require metrics to clearly distinguish the good choices from the less suitable ones. For some decisions, e.g., selecting the modeling approach, providing suitable metrics is still an open challenge. Knowledge about constraints on applying one method or the other, and interdependencies and implications of using one or the other method on future activities will reduce uncertainties in the overall process.

Within the engineering of models, well-founded answers to the questions of which step to do next and which method to use largely determine the efficiency and effectiveness of the model engineering process. Referring to the first question, and for orchestrating the diverse processes that are involved in modeling engineering, workflow-based approaches might be exploited to make these processes explicit and traceable. These approaches facilitate evaluation of different phases of model life cycle engineering, including validation and verification of models, and, thus, add to the credibility of M&S. However, this requires a high degree of standardization of these processes. This might be achievable for specific subprocesses of validation or verification, e.g., how to execute and analyze a parameter scan given a specific model. However, the overall process of a simulation study is highly interactive and thus one might only be able to define general constraints on the engineered artifacts, e.g., if the conceptual model (if we interpret conceptual model as a representation of requirements or invariants that refer to the simulation model) changes, so does the stage of the process model, requiring a new validation phase.

3.3.2 Effectiveness Measures

In a model-based engineering (MBE) approach, the development team evolves a set of models throughout the system life cycle to support design, analysis, and verification of the system under development. These models are intended to express aggregate knowledge about the system to enable communications and shared understanding among the development team and other program stakeholders. Program leadership must continue to determine what knowledge must be acquired at any given point in the life cycle to maximize the likelihood of program success. The type of knowledge to be acquired can help identify the kind of design and analysis models that should be further developed and updated.

This knowledge can be acquired by performing engineering tasks that involve different kinds of models, such as performing a trade study to select among alternative system architectures, performing an analysis to determine a system error budget, updating electrical, mechanical, or software designs, or analyzing a particular design for reliability, safety, manufacturability, or maintainability. Determining what knowledge is needed becomes more challenging as the complexity of the system increases, and as the complexity of the organization that develops the system increases (e.g., large geographically distributed teams).

The research challenge is to define one or more effectiveness measures that can guide the knowledge acquisition process and associated model development and evolution throughout the system life cycle. In other words, how do you determine the additional knowledge at each point in time that provides best value to the program stakeholders? The research can benefit from data that has been collected over many years to find a solution. For example, the following figures are typical examples of trends that indicate the impact of collecting certain kinds of knowledge on the overall cost of system development.

In Fig. 3.1 the lower curves reflect the percentage of the total life cycle cost that is expended as a function of the phase of the program life cycle. As indicated, much of the cost is expended in the later life cycle phases. However, as shown in the upper curve, the percentage of the life cycle cost that is committed occurs much earlier in the life cycle. This finding shows the importance of early design decisions based on the available knowledge. The cost to fix a defect increases exponentially as a function of the phase in the product life cycle where the defect is detected (Boehm 1981; McGraw 2006). Acquiring the knowledge to surface defects early can substantially reduce the total system life cycle cost.

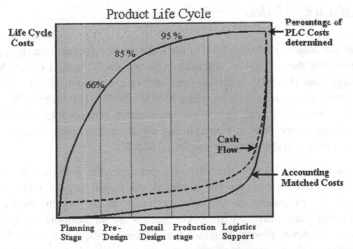

Adapted from the CAM-I conceptual design p. 140. Original source, Blanchard, Design and Manage to Life Cycle.

Fig. 3.1 Committed and actual lifecycle costs (Berliner and Brimson 1988) citing (Blanchard 1978)

The following are some suggested factors to be considered for research:

- Aggregate knowledge goals at particular milestones in a system development life cycle.
- Knowledge elements that contribute to the aggregate knowledge.
- Knowledge elements associated with different aspects of the system of interest and its environment.
- A value function associated with acquiring knowledge elements at each point in time, and its impact on the probability of program success.
- Cost to acquire the knowledge elements at a given point in the life cycle.
- Cost associated with acquiring incorrect knowledge at a given point in the life cycle.
- Relationship between the effectiveness measure (value vs. cost) and more traditional risk measures.

The acquisition of knowledge across a life cycle can be thought of as a trajectory whose aim is to maximize program success. The value function of acquired knowledge is dependent on both the knowledge elements and the sequence in which these elements are acquired, since there are dependencies among the knowledge elements. For example, during the concept phase of a vehicle's development, it is often important to acquire knowledge about vehicle sizing and system level functionality to meet mission performance requirements, but it may not be important to acquire knowledge about the detailed software design algorithms.

3.3.3 Maturity Models

The Capability Maturity Model (CMM) for software development has played a key role to guarantee the success of software projects (Paulk et al. 1993). CMM and CMM Integration (CMMI) originated in software engineering, but have been applied to many other areas over the years (CMMI 2016a). However, in M&S, there is no such standardized and systematic assessment methodology developed for M&S processes. Some related research and development results can be used as references to establish the maturity model of M&S:

- Software life cycle models describe core processes for software development. Following proven processes for model development begins with an understanding and execution of the core activities in an organization's chosen development path. Software life cycle models are an example of core proven processes for development. Whether the life cycle model chosen is the classic waterfall model or more modern iterative versions, all have aspects of requirements development, design, implementation, integration, test, etc., utilized in a way that best fits the size of the organization, the size of the project, or the constraints of the customer and developer.

- Software CMMI or CMMI for Development shows the success of maturity for general software development. CMMI was originally developed at Carnegie Mellon University and the federally funded Software Engineering Institute (SEI). The CMMI Institute reports that thousands of CMMI appraisals are completed every year in dozens of countries (CMMI 2016b). CMMI enables organizations to be viewed and certified as being mature and capable of carrying out intended activities to a certain level or degree of expertise, which lends that degree of credibility to the components developed by those activities. CMMI assigns *capability* levels to process improvement achievement in individual process areas. Therefore, a certain part of an organization may be identified or certified at a level 3 out of 4 for Configuration Management, but capability level 2 out of 4 for Maintenance. CMMI assigns *maturity* levels to process improvement achievement in multiple process areas or a set of process areas and applies to the scope of the organization that was evaluated/certified such as a department, a division, or the entire company—Level 5 being the highest achievable level of maturity.
- The Federation Development and Execution Process (FEDEP) describes core processes for simulation development. FEDEP was initially released in 1996 as the first common process for the development of simulations and was specifically for guidance in creating High Level Architecture (HLA) federations (IEEE Standards Association 2010). These common methodologies and procedures consisted of six steps: (1) Define Objectives, (2) Develop Conceptual Model, , (3) Design Federation, (4) Develop Federation, (5) Integrate and Test, and (6) Execute and Prepare Results. These six steps included specific work products that were inputs and outputs to each step. These steps and the FEDEP process paralleled the software development process and could serve as the initial draft of the core processes for model and simulation development that would be the basis for an examination of an organization's ability to robustly, reliably, and repeatedly develop credible models and simulations, i.e., identify the capability and maturity of organizations or portions of organizations.

System of Systems describes corollaries that may exist within Systems Engineering (Zeigler and Sarjoughian 2013).

- DoDAF describes the notion of multiple views of simulations (US Department of Defense 2016b).

Taking CMM/CMMI as a basis, a capability maturity model for modeling and simulation process (MS-CMMI) could be established by:

- Finding the differences and similarities between the processes of modeling/simulation and software development by analyzing characteristics of modeling process and simulation of complex systems, then define indicators and metrics for M&S processes.

- Setting up a MS-CMMI evaluation system (evaluation methods, standards, tools, organizations, etc.) to assess the structured level of capabilities of model developers or model users (use the model to do simulation).

Achieving these goals requires research in:

- Quantitative analysis of the complexity and uncertainties in modeling processes.
- Optimization of modeling processes.
- Risk analysis and control of modeling processes.
- Quantitative measurement of model life cycle quality and cost.
- Notional mappings with CMMI, etc.
- Identification and description of processes and work products necessary at differing levels of the Modeling and Simulation Maturity Model, when and why they are needed and who performs them.

3.3.4 Validation

As simulation models become more complex, validation of conceptual models and understanding their role in the broader process of validation will continue to be important research areas. Of course, understanding validation of conceptual models is dependent on a precise definition of the terms "conceptual model" and "validation." This section argues that a better consensus is needed on the first term, while a careful review of the validation literature will reveal the same for the second.

This is particularly apparent across M&S communities of practice. For example, the training and engineering communities intersect the broader M&S community, but M&S stakeholders in those communities draw heavily from skill sets based in different scientific disciples and different perspectives of the role of modeling and simulation. The M&S community's challenge is to address universally applicable concepts, like conceptual models in validation, from a holistic perspective with theory that is satisfying to all the stakeholders and technology that is germane to a broad set of problems (in the case of the stated example, simulation theory and technology that is useful for both the social scientist and the engineer).

Consider a few simple questions. What does it mean to validate a conceptual model? How does a conceptual model that is suitable for a specific use inform the development of other simulation process artifacts? How do the various stakeholders in the simulation activity use the conceptual model, valid or otherwise? Some researchers will see these as easily answered in their particular domains, but will find their conclusions quite different between domains. So, for the discussion in this section we consider terminology in the broadest context possible.

Consider modeling paradigms as equivalence classes on the set of conceptual models. Each paradigm has defining characteristics in terms of conceptual modeling language or formalism. These characteristics define every instance of a conceptual model as belonging to one class or another, or perhaps none. For well-developed theory, further properties and theorems will follow to enable reasoning on all of the elements of each class in general without resorting to building, coding, and executing every instance to understand its properties. As we develop more rigorous and explicit conceptual models that bridge between the referent system and the computer simulation, methods for validation will become even more critical.

Simulation frameworks that include a category for conceptual models permit the side-by-side comparison to and facilitate discussion of related artifacts. For example, Balci and Ormsby (2007), Petty (2009), and Sargent (2013) provide frameworks that include conceptual models in this context. Advances in conceptual modeling will drive the need for new frameworks to explain the properties of conceptual models, and the relations between them, the referent system, and the computer simulation.

Some researchers would consider that the conceptual model is an appropriate artifact to analyze for suitability for use. Although recent work in validation theory is looking hard at the implication of risk in the decision to use particular kinds of simulation, and propagation of error in simulation, more basic research is needed to develop a robust model-based decision theory. Accuracy is well understood, particularly in the context of physics-based models, but its use in simulation is not well defined. When deciding on the kind of simulation to inform a particular decision, acceptability criteria are often subjective and little theory exists to objectify the decision analysis. A well-developed model-based decision theory will recast validation in the language of decision theory, defining use in a rigorous way, clearly differentiating objective from subjective elements of the use decision, and providing a defensible basis for using models and simulations to inform decision-making (Weisel 2012).

The logical next step for advances in theory involving validation of conceptual models is to incorporate these advances in simulation development environments. As model life cycle engineering develops, tools for validation of conceptual models are needed to keep pace. As conceptual modeling languages and formalisms become useful additions to simulation development environments, tools using well-defined conceptual models within the broader process of validation will improve the quality and defensibility of the simulation end product.

It is well understood that validation is best considered early in the development process (see the previous two subsections)—there should be no difference when considering conceptual models in the mix. As development environments would benefit from rigorous application of conceptual models in the development process, so too would the consideration of validation from the earliest life cycle stages. New technologies and tools are needed to incorporate validation of conceptual models throughout the simulation life cycle.

3.4 Conceptual Model Architecture and Services

Many modeling paradigms exist for most kinds of domain problems, applied to knowledge from many engineering disciplines. Understanding complex systems requires integrating these into a common composable reasoning scheme (NATO Research and Technology Organization 2014). The software and the system engineering communities have overcome similar challenges using architecture frameworks (e.g., OMG's Unified Architecture Framework OMG 2016), but modeling and simulation does not have a similarly mature integration framework. The first subsection below concerns architectures for conceptual modeling, while the second outlines infrastructure services needed to support those architectures.

3.4.1 Model Architecture

At the foundation of a modeling architecture should be a fundamental theory of models, to enable reusability, composability, and extensibility. What theory of models could support the implementation of a model architecture? An epistemic study of existing modeling and integration paradigms is necessary to develop a theory of models. This should include a taxonomy of modeling paradigms, semantics, syntaxes, and their decomposition into primitives that operate under common rules across paradigms, to integrate them as required by complex systems.

Model architecture is needed to unify different classes of models developed using different paradigms. An architecture is the glue specifying interfaces, rules of operation, and properties common across modeling paradigms, enabling models to be interconnected at multiple levels of conceptual abstraction. What is meaningful to connect? What is not? An architecture goes far beyond conventional model transformations and gateways, though these are also essential to comprehension of multiparadigm modeling processes. An architecture is about persistent coexistence and coevolution in multiple domains at multiple levels of abstraction. How can a model architecture framework connect models that operate according to different sets of laws? For example, critical infrastructure protection requires connecting country, power grid, internet, economy, command, and control, etc. Combat vehicle survivability requires connecting humans, materials, optics, electromagnetics, acoustics, cyber, etc. What mechanisms are required to efficiently interact between different sets of laws (e.g., layered architecture)? What level of detail is required to observe emerging behaviors between different sets of laws when integrated? How should a model architecture be implemented, in which format, using which tools? As a model architecture matures, successful design patterns should emerge for the most common reusable interconnections between disciplines. What are these design patterns in each community of interest?

Model architecture sets the rules to meaningfully interconnect models from different domains. Generalizing and publishing rules for widespread modeling paradigms would allow composing and reusing models that comply with the architecture and complex system simulations will become achievable. As an example of interconnected models across domains, start with a Computer-Aided Design (CAD) model representing a physical 3D object in terms of nodes and facets. In the CAD paradigm, objects can be merged to interconnect. A related Finite Element Model (FEM) represents continuous differential equations for physical laws between boundary layers. It can be used to compute the fluid dynamics during combustion. FEM models can interconnect at the level of physical laws to compute the temperature distribution from the combustion products distribution for instance. They also interconnect with a CAD model at the mesh level. A computer graphics model enables display of objects as seen from particular viewpoints. It interconnects with CAD and FEM models to map materials and temperature to facets for the purpose of generating an infrared scene image in the field of view of a sensor. A functional model of a surveillance system can represent discrete events involved in changing a sensor mode as a function of the mission. The functional model interconnects with the computer graphics model at the sensor parameter level. Finally, a business process model can represent a commander's mission planning. It can interconnect with a functional model by changing the mission.

Figures of merit must be developed to demonstrate how well a model architecture facilitates composition of multiparadigm, multiphysics, multiresolution models. The performance of a model architecture must be checked against interdisciplinary requirements using metrics for meaningfulness and consistency. How can we test a particular integration for validity? How can it be done efficiently over large-scale complex simulations? How can it be done by a non-expert? What mechanisms should a model architecture framework include to support checking for conceptual consistency?

Integration complexity and coupling between the degrees of freedom of individual components and the degrees of freedom of the integration are yet to be understood. When integrating a model in a complex simulation, what details can be ignored and still ensure a valid use of that model? What details cannot be ignored?

Reliable model integration depends on sufficient formality in the languages used, as described in Sect. 3.1. In particular, formal conceptual models of both the system of interest (referent) and analysis provide a basis for automating much of analysis model creation through model-to-model transformation. As an example, consider the design of a mechanical part or an integrated circuit. The CAD tools for specifying these referents use a standard representation, with a formal semantics and syntax. For particular kinds of analyses—such as response in an integrated circuit—simulations are essentially available at the push of a button. Formalism in the specification of the referent enables automation of certain analyses. This pattern is well demonstrated, e.g., in the use of Business Process Modeling Notation (BPMN) to define a business process, and then automating the translation of this model into a hardware/software implementation specification. The Object Management Group has developed standard languages for model-to-model transformations. At present, there are only limited demonstrations of applying this approach to systems modeling. Automating this kind of model-to-model transformation captures knowledge about how to create analysis models from referent models, so perhaps the most fundamental question is: where should this knowledge reside—should it be captured in the referent modeling language, in the analysis modeling language, in the transformation, or perhaps spread throughout? Formalization of mappings between conceptual models of a referent and its analysis models is critical to building reliable bridges between descriptions of the referent and specifications of a simulation model and its computational implementation.

3.4.2 Services

The success of large-scale integration of knowledge required by complex systems fundamentally depends on modeling and simulation infrastructure services aggregated into platforms. These enable affordable solutions based on reusing domain-specific models and simulators, as well as integrating them into a multi-model cosimulation. For example, understanding vulnerabilities and resilience of complex engineered systems such as vehicles, manufacturing plants, or electric distribution networks requires the modeling and simulation-based analysis of not only the abstracted dynamics, but also some of the implementation details of networked embedded control systems. Systems of such complexity are too expensive to model and analyze without reuse and synergies between projects.

Services need to enable open model architecture development and sharing of model elements at all levels. How can a common conceptual modeling enterprise be launched involving many stakeholders? How can a conceptual model be augmented with knowledge from different contributors (e.g., wiki)? How does it need to be managed? What structure should the conceptual model have? What base ontologies are required (e.g., ontology of physics)? How can conceptual model components be implemented in executable model repositories and how can components plug and play into simulation architectures? Guiding principles must also be defined and advertised. What guidance should modelers follow to be ready for a collaborative conceptual modeling enterprise in the future? Standard theory of models, architecture, design patterns, consistency tests, modeling processes, and tools will arise naturally as the modeling science matures.

Services can be aggregated into three horizontal integration platforms:

- In *Model Integration Platforms*, the key challenge is to understand and model interactions among a wide range of heterogeneous domain models in a semantically sound manner. One of the major challenges is semantic heterogeneity of the constituent systems and the specification of integration models. Model integration languages have become an important tool for integrating complex, multimodeling design automation and simulation environments. The key idea is to derive opportunistically an integration language that captures only the cross-domain interactions among (possibly highly complex) domain models (Cheng et al. 2015).

- *Simulation Integration Platforms* for cosimulation have several well-established architectures. The High Level Architecture (HLA) (IEEE Standards Association 2010) is a standardized architecture for distributed computer simulation systems. The Functional Mockup Interface (Modelica Association 2014a) for cosimulation is a relatively new standard targeting the integration of different simulators. In spite of the maturity and acceptance of these standards, there are many open research issues related to scaling, composition, large range of required time resolution, hardware-in-the-loop simulators and increasing automation in simulation integration.

- *Execution Integration Platforms* for distributed cosimulations are shifting toward cloud-based deployment, developing simulation as a service using models via web interfaces and increasing automation in dynamic provisioning of resources as required. More will be said about this in the next chapter.

References

Adiga, A., C. Kuhlman, M. Marathe, S. Ravi, D. Rosenkrantz, and R. Stearns. 2016. Inferring local transition functions of discrete dynamical systems from observations of system behavior. *Theoretical Computer Science*.

Baader, F., D. Calvanese, D. McGuinness, D. Nardi, and P. Patel-Schneider (eds.). 2010. The description logic handbook: Theory, implementation and applications, 2nd ed.

Balci, O. and W. F. Ormsby. 2007. Conceptual modelling for designing large-scale simulations. *Journal of Simulation* 1: 175–186.

Barrett, C., S. Eubank, V. Kumar, and M. Marathe. 2004. Understanding large scale social and infrastructure networks: a simulation based approach. SIAM news in Math Awareness Month on The Mathematics of Networks.

Barrett, C., S. Eubank, and M. Marathe. 2006. Modeling and simulation of large biological, information and socio-technical systems: an interaction based approach. *Interactive computation*, 353–394. Berlin Heidelberg: Springer.

Batarseh, O. and L. McGinnis. 2012. System modeling in SysML and system analysis in Arena. In *Proceedings of the 2012 Winter Simulation Conference*.

Berliner, C. and J. Brimson, (eds.). 1988. Cost management for today's advanced manufacturing: the CAM-I conceptual design. Harvard Business School Press.

BioNetGen. 2016. "BioNetWiki." Accessed 29 Aug 2016. http://bionetgen.org.

Blanchard, B. 1978. *Design and manage to life cycle cost*. Dilithium Press.

Bock, C., M. Gruninger, D. Libes, J. Lubell, and E. Subrahmanian. 2006. *Evaluating reasoning systems*. Report: U.S National Institute of Standards and Technology Interagency. 7310.

Bock, C., and J. Odell. 2011. Ontological behavior modeling. *Journal of Object Technology* 10 (3): 1–36.

Bock, C., M. Elaasar. 2016. Reusing metamodels and notation with Diagram Definition. *Journal of Software and Systems Modeling*.

Boehm, B. 1981. *Software engineering economics*. Prentice-Hall.

Bouissou, O. and A. Chapoutot. 2012. An operational semantics for Simulink's simulation engine. Languages, compilers, tools and theory for embedded systems. In *Proceedings of the 13th ACM SIGPLAN/SIGBED International Conference, (LCTES'12)*. 129–138.

Cellier, F. 1991. *Continuous systems modelling*. Springer.

Cheng, B., T. Degueule, C. Atkinson, S. Clarke, U. Frank, P. Mosterman, and J. Sztipanovits. 2015. Motivating use cases for the globalization of CPS. In *Globalizing Domain-Specific Languages*, LNCS, vol. 9400, 21–43. Springer.

CMMI Institute. 2016a. "CMMI Models." Accessed 29 Aug 2016. http://cmmiinstitute.com/cmmi-models.

CMMI Institute. 2016b. 2015 Annual Report to Partners, Accessed 6 Sept 2016. http://partners.cmmiinstitute.com/wp-content/uploads/2016/02/Annual-Report-to-Partners-2015.pdf.

Eclipse Foundation. 2016a. Eclipse modeling framework. Accessed 29 Aug 2016. https://eclipse.org/modeling/emf.

Eclipse Foundation. 2016b. Graphical modeling project. Accessed 29 Aug 2016. http://www.eclipse.org/modeling/gmp.

Emerson, M., S. Neema, and J. Sztipanovits. 2006. *Metamodeling languages and metaprogrammable tools*. Handbook of Real-Time and Embedded Systems: CRC Press.

Eskridge, T., and R. Hoffman. 2012. Ontology creation as a sensemaking activity. *IEEE Intelligent Systems* 27 (5): 58–65.

Finney, A., M. Hucka, B. Bornstein, S. Keating, B. Shapiro, J. Matthews, B. Kovitz, M. Schilstra, A. Funahashi, J. Doyle, and H. Kitano. 2006. Software infrastructure for effective communication and reuse of computational models. In *System Modeling in Cell Biology: From Concepts to Nuts and Bolts*. Massachsetts Institute of Technology Press.

Flatscher, R. 2002. Metamodeling in EIA/CDIF—meta-metamodel and metamodels. *ACM Transactions on Modeling and Computer Simulation*. 12 (4): 322–342.

Gašević́a, D., and V. Devedžić. 2006. Petri net ontology. *Knowledge-Based Systems* 19 (4): 220–234.

Harrison, N., W.F. Waite. 2012. Simulation conceptual modeling tutorial. In *Summer Computer Simulation Conference*, July 2012.

Harvard Medical School. 2016. KaSim: A rule-based language for modeling protein interaction networks. Accessed 29 Aug 2016. http://dev.executableknowledge.org.

Helms, T., C. Maus, F. Haack, and A. M. Uhrmacher. 2014. Multi-level modeling and simulation of cell biological systems with ML-rules: a tutorial. In *Proceedings of the Winter Simulation Conference*, 177–191.

Helms, T., R. Ewald, S. Rybacki, A. Uhrmacher. 2015. Automatic runtime adaptation for component-based simulation algorithms. *ACM Transactions on Modeling and Computer Simulation*, 26 (1).

Huang, E., Ky Sang Kwon, and L. F. McGinnis. 2008. Toward on-demand wafer fab simulation using formal structure and behavior models. In *Proceedings of the 2008 Winter Simulation Conference*.

IEEE Standards Association. 2010. 1516–2010-IEEE Standard for Modeling and Simulation (M&S) High Level Architecture (HLA). http://standards.ieee.org/findstds/standard/1516-2010.html.

Institute for Software Integrated Systems. 2016. "WebGME." Accessed 29 Aug 2016. https://webgme.org.

Jackson, E., and J. Sztipanovits. 2009. Formalizing the structural semantics of domain-specific modeling languages. *Journal of Software and Systems Modeling* 8: 451–478.

Karsai, G., M. Maroti, A. Lédeczi, J. Gray, and J. Sztipanovits. 2004. Composition and cloning in modeling and meta-modeling. *IEEE Transactions on Control System Technology*. 12 (2): 263–278.

Le Novère, N., M. Hucka, H. Mi, S. Moodie, F. Schreiber, A. Sorokin, E. Demir, K. Wegner. M. Aladjem, S. Wimalaratne, F. Bergman, R. Gauges, P.Ghazal, H. Kawaji, L. Li, Y. Matsuoka, A. Villéger, S. Boyd, L. Calzone, M. Courtot, U. Dogrusoz, T. Freeman, A. Funahashi, S. Ghosh, A. Jouraku, A, S. Kim, F. Kolpakov, A. Luna, S. Sahle, E. Schmidt, S. Watterson, S., G. Wu, I. Goryanin, D. Kell, C. Sander, H. Sauro, J. Snoep, K. Kohn, and H. Kitano. 2009. The systems biology graphical notation. *Natural Biotechnolology*. 27 (8): 735–741.

Mannadiar, R. and H. Vangheluwe. 2010. Domain-specific engineering of domain-specific languages. In *Proceedings of the 10th Workshop on Domain-Specific Modeling*.

McGinnis, L. and V. Ustun. 2009. A simple example of SysML driven simulation. In *Proceedings of the 2009 Winter Simulation Conference*.

McGraw, G. 2006. *Software Security: Building Security In*. Addison-Wesley.

Modelica Association. 2014a. Functional markup interface. https://www.fmi-standard.org/downloads#version2.

Modelica Association. 2014b. Modelica® -a unified object-oriented language for systems modeling, language specification, version 3.3, revision 1. Accessed 6 Sept 2016. https://www.modelica.org/ documents/ModelicaSpec33Revision1.pdf.

Mortveit, H., and C. Reidys. 2008. *An Introduction to Sequential Dynamical Systems*. Springer.

NATO Research and Technology Organisation. 2012. Conceptual Modeling (CM) for Military Modeling and Simulation (M&S). Technical Report TR-MSG-058, https://www.sto.nato.int/publications/STO%20Technical%20Reports/RTO-TR-MSG-058/$$TR-MSG-058-ALL.pdf, July 2012.

Novak, J. 1990. Concept maps and Vee diagrams: two metacognitive tools for science and mathematics education. *Instructional Science* 19: 29–52.

Object Management Group. 2013. Business process model and notation. http://www.omg.org/spec/BPMN.

Object Management Group. 2014. Object constraint language. http://www.omg.org/spec/OCL.

Object Management Group. 2015a. Systems modeling language. http://www.omg.org/spec/SysML.

Object Management Group. 2015b. Unified modeling language. http://www.omg.org/spec/UML.

Object Management Group. 2013/2016. Unified architecture framework. http://www.omg.org/spec/UPDM, http://www.omg.org/spec/UAF.

Paulk, M., W. Curtis, M. Chrissis, and C. Weber. 1993. Capability maturity model, version 1.1. Technical Report Carnegie Mellon University Software Engineering Institute. CMU/SEI-93-TR-024 ESC-TR-93-177, February 1993.

Petty, M. D. 2009. Verification and validation. In *Principles of Modeling and Simulation: A Multidisciplinary Approach*. John Wiley & Sons, 121–149.

Quillian, M. 1968. Semantic memories. In *Semantic Information Processing*. Massachusetts Institute of Technology Press, 216–270.

Reichgelt, H. 1991. *Knowledge Representation: An AI Perspective*. Ablex Publishing Corporation.

Reinhartz-Berger, I., and D. Dori. 2005. OPM vs. UML–experimenting with comprehension and construction of web application models. *Empirical Software Engineering*. 10 (1): 57–80.

Robinson, S. 2013. Conceptual modeling for simulation. In *Proceedings of the 2013 Winter Simulation Conference*.

Rosenkrantz, D., M. Marathe, H. Hunt III, S. Ravi, and R. E. Stearns. 2015. Analysis problems for graphical dynamical systems: a unified approach through graph predicates. In *Proceedings of the 2015 International Conference on Autonomous Agents and Multiagent Systems (AAMAS'15)*. *International Foundation for Autonomous Agents and Multiagent Systems*, Richland, SC, 1501–1509.

Sargent, R.G. 2013. Verification and validation of simulation models. *Journal of Simulation* 7: 12–24.

Sattler, U., D. Calvanese, and R. Molitor. 2010. Relationships with other formalisms, 149–193 (Baader et al. 2010).

Simko, G., D. Lindecker, T. Levendovszky, S. Neema, and J. Sztipanovits. 2013. Specification of cyber-physical components with formal semantics–integration and composition. In *Model-Driven Engineering Languages and Systems*. Springer Berlin Heidelberg, 471–487.

Sprock, T. and L. F. McGinnis. 2014. Simulation model generation of Discrete Event Logistics Systems (DELS) using software patterns. In *Proceedings of the 2014 Winter Simulation Conference*.

Thiers, G. and L. McGinnis. 2011. Logistics systems modeling and simulation. In *Proceedings of the 2011 Winter Simulation Conference*.

U.K. Ministry of Defense. 2016. MOD architecture framework. Accessed 29 Aug 2016. https://www.gov.uk/guidance/mod-architecture-framework.

U.S. Department of Defense. 2016a. The DoDAF architecture framework version 2.02. Accessed 29 Aug 2016. http://dodcio.defense.gov/Library/DoD-Architecture-Framework.

U.S. Department of Defense. 2016b. DoDAF viewpoints and models. Accessed 29 Aug 2016. http://dodcio.defense.gov/Library/DoD-Architecture-Framework/dodaf20_viewpoints.

Warnke, T., T. Helms, and A. M. Uhrmacher. 2015. Syntax and semantics of a multi-level modeling language. In *Proceedings of the 3rd ACM SIGSIM Conference on Principles of Advanced Discrete Simulation*, 133–144.

Weisel, E. W. 2012. A decision-theoretic approach to defining use for computer simulation. In *Proceedings of the 2012 Autumn Simulation Multi-Conference*.

Zeigler, B., H. Praehofer, and T. Kim. 2000. *Theory of Modeling and Simulation*, 2nd ed. Academic Press.

Zeigler, B., and S. Sarjoughian. 2013. *Guide to Modeling and Simulation of Systems of Systems*. London: Springer.

Chapter 4
Computational Challenges in Modeling and Simulation

Christopher Carothers, Alois Ferscha, Richard Fujimoto,
David Jefferson, Margaret Loper, Madhav Marathe,
Pieter Mosterman, Simon J.E. Taylor and Hamid Vakilzadian

Computational algorithms and software play a central role in all computer models and simulations. A computer simulation can be viewed as a collection of state variables and data structures that represent the state of the system under investigation and algorithms that transform that state to capture the evolution of the system state

---

The original version of this chapter was revised: Contributor name and corresponding affiliation have been included. The erratum to this chapter is available at https://doi.org/10.1007/978-3-319-58544-4_7

C. Carothers
Rensselaer Polytechnic Institute, Troy, NY, USA
e-mail: chris.carothers@gmail.com

A. Ferscha
Johannes Kepler Universität Linz, Linz, Austria
e-mail: ferscha@soft.uni-linz.ac.at

R. Fujimoto (✉)
Georgia Institute of Technology, Atlanta, GA, USA
e-mail: fujimoto@cc.gatech.edu

D. Jefferson
Lawrence Livermore National Laboratory, Livermore, CA, USA
e-mail: jefferson6@llnl.gov

M. Loper
Georgia Tech Research Institute, Atlanta, GA, USA
e-mail: Margaret.Loper@gtri.gatech.edu

M. Marathe
Virginia Tech, Blacksburg, VA, USA
e-mail: mmarathe@vbi.vt.edu

P. Mosterman
MathWorks, Natick, MA, USA
e-mail: Pieter.Mosterman@mathworks.com

© Springer International Publishing AG (outside the USA) 2017
R. Fujimoto et al. (eds.), *Research Challenges in Modeling
and Simulation for Engineering Complex Systems*, Simulation Foundations,
Methods and Applications, DOI 10.1007/978-3-319-58544-4_4

45

over time. The algorithms encode the rules that govern the behavior of the system. In many cases, the basis for these behaviors is specified through mathematics, e.g., differential equations derived from physical laws. In other simulation models, the behaviors are specified in logical rules that encode the causal relationships among the components making up the system. These computational rules may be used to determine the new state of the system in the next "clock tick" or time step of the simulation computation. In other simulations, the changes in system state may occur at irregular points in simulation time, governed by the occurrence of "interesting" events such as a doctor finishing a consultation with a patient or a machine finishing the processing of a part in a manufacturing system. Regardless, computational methods and software are critical elements in modeling and simulation.

New challenges are arising in computational methods for modeling and simulation that create new research problems that must be addressed. This is because application requirements are changing on the one hand, and the underlying computational platforms are changing on the other. For example, reliable simulation models are essential to determine the impact of new policies and technologies on the evolution of cities, an area of increased interest with phenomena such as global warming creating new challenges. The infrastructures making up a city such as water, transportation, and energy are highly dependent on each other. For example, electrification of the vehicle fleet will clearly have a direct impact on vehicle emissions. But electrification has other impacts as well. The demand for electricity in households will increase, which in turn impacts the emissions produced by power generation plants as well as the amount of water they consume. In some cases, this is the same water used for food production, resulting in other impacts on the economy. When one considers other emerging technologies such as household generation of power through solar panels and more broadly smart homes, the introduction of automated vehicles, commercial use of drones for package delivery, the introduction of smart electrical power grids, and changing human behaviors resulting from these innovations, the emerging phenomena resulting from the confluence of these interactions are not well understood.

At the same time, the computing platforms on which simulations execute are undergoing a different kind of revolution. For decades, the performance of computer hardware doubled every 18 months in accordance with Moore's Law. These improvements derived largely from increases in the clock rate used to drive computer circuits. These improvements in clock speed stopped around 2004 due to an inability to dissipate heat from these circuits as they were clocked at a higher rate. Now, advances in hardware performance are being derived almost entirely from

S.J.E. Taylor
Brunel University London, Uxbridge, UK
e-mail: Simon.Taylor@brunel.ac.uk

H. Vakilzadian
University of Nebraska-Lincoln, Lincoln, NE, USA
e-mail: hvakilzadian@unl.edu

exploitation of parallel processing. The number of processors or cores in computing devices has been increasing rapidly across all platforms, from high-performance supercomputers down to computers in handheld devices such as smartphones. Another related phenomenon dramatically changing the hardware landscape is the emergence graphical processing unit (GPU) devices for a much broader range of application than rendering graphics, for which they were originally designed. The high-volume production of these devices has lowered their cost, making them increasingly attractive for computationally demanding tasks. A third major hardware trend is the explosion of mobile computing devices that continue to increase in power and sophistication. These hardware changes have great implications in the development of computational algorithms for computer simulations, which are among the most computation-intensive applications that exist. There are many research challenges that call for the development of novel computational methods, as will be discussed later in this chapter.

Other major trends in computing include cloud computing, "big data," and the Internet of Things. Each of these developments has major ramifications in modeling and simulation. Cloud computing provides a platform that can make access to high-performance computing facilities as straightforward as having access to the Internet, opening broader opportunities for exploitation of computation-intensive simulations. Modeling and simulation has long utilized data analysis techniques for tasks such as characterizing inputs and specifying relevant parameters for simulation models. Big data technologies offer new capabilities that simulations can readily exploit. While big data and advances in artificial intelligence are creating unprecedented capabilities for situation awareness, i.e., characterizing and interpreting the state of operational systems, modeling and simulation offers a predicative capability that cannot be achieved through data analysis algorithms alone. In addition, the Internet of Things creates many new rich sources of data that are again synergistic to modeling and simulation offering unprecedented opportunities for modeling and simulation to be embedded in the real world and to have enormous impacts in society. These emerging platforms and computation technologies offer exciting new opportunities to increase the impact of modeling and simulation in the context of managing operational systems. Lastly, dynamic data-driven application systems (Darema 2004), a paradigm that encompasses real-time data driving computations and simulations, are used in a feedback loop to enhance monitoring and/or aid in the management of operational systems.

This chapter describes important computational challenges that must be addressed for modeling and simulation to achieve its fullest potential to address the new requirements of contemporary applications and to maximally exploit emerging computing platforms and paradigms. The first section of this chapter focuses on emerging computing platforms and computational challenges that must be addressed to effectively exploit them. These range from massively parallel simulations executing on supercomputers containing millions of cores, to effectively exploiting new platforms with heterogeneous computing elements such as GPU accelerators, to field programmable gate arrays (FPGAs), cloud computing environments, and mobile computing platforms. Radical, new computing approaches such as neuromorphic computing that are loosely modeled on the human brain are also discussed. The section that follows focuses on challenges arising where simulations become pervasive and appear everywhere utilizing paradigms such as DDDAS mentioned earlier and cyber-physical systems. Creating, understanding, and managing large-scale distributed systems of simulations interacting with each other to manage operational systems and subsystems present major challenges and raise important concerns in privacy, security, and trust. Research isrequired both to identify fundamental principles concerning such simulations and to establish a theory behind their operation.

The third section raises the question of how the modeling and simulation community should manage the plethora of models that already exists, and continues to expand as new modeling approaches are developed. Complex systems often involve many subsystems, each of which may be a complex system in its own right. Understanding systems such as these will inevitably require a host of different modeling approaches to be integrated, including not only different types of models, but models operating across vastly different scales in time and space. The relationship among these different modeling approaches is poorly understood, as is determining computational methods to best combine them to address key questions in large, heterogeneous, complex systems. Are there underlying theories that can be used to combine traditionally distinct areas such as continuous and discrete event simulation? What computational methods and algorithms are required to successfully exploit this plethora of models? Many simulation trials will be required in any study. Are there new techniques to improve the execution of the so-called ensemble simulations?

The section that follows explores the relationship between modeling and simulation and big data, and highlights the synergies that naturally arise between these technologies. Simulation analytics represents a new paradigm expanding and exploiting these synergies. Key research questions concerning model and data representation, challenges in managing large-scale data and live data streams, and an approach termed qualitative modeling are discussed.

In summary, there are numerous computational challenges that must be addressed for modeling and simulation to achieve maximal impact in light of new application requirements and emerging hardware and software computing platforms. This chapter highlights areas where advances are required to maximize the effectiveness and impact of modeling and simulation in society.

4.1   Exploiting Emerging Computing Platforms

The general computing architectures used for most large-scale simulations have been similar for the last 30 years: shared memory multicore or multiprocessor systems and tightly coupled distributed memory clusters. But computing platforms have undergone dramatic changes in the last decade, changes that are only modestly exploited by modeling and simulation technologies today. Examples of computing platforms requiring greater attention for modeling and simulation applications include massively parallel supercomputers, heterogeneous computing systems including graphical processing unit (GPU) accelerators, and field programmable gate arrays. The growing, widespread adoption of cloud and mobile computing create new opportunities and challenges for modeling and simulation. Effective exploitation of these platforms requires careful consideration of how simulation computations can best exploit and operate under the constraints imposed by the underlying platform while meeting execution time and energy consumption goals for contemporary applications. Research challenges for each of these new, emerging computing platforms are discussed next, and discussed in greater detail in Fujimoto (2016).

4.1.1 Massively Parallel Simulations

The number of processors (cores) in the most powerful supercomputers has exploded in the last decade. While the number of processors in the most powerful machines remained relatively stable, ranging from a few thousand in 1995 to ten thousand in 2004, this number began increasing dramatically in 2005. In November 2015, the Tianhe-2 machine, rated the most powerful supercomputer in the world, contained over 3 million cores. Effective exploitation of the computing power provided by these machines for large-scale simulation problems requires a paradigm shift in the modeling and simulation community.

This trend is exemplified by experimental data reported in the literature in the parallel discrete event simulation field (Barnes et al. 2013). Performance measurements of telecommunication network simulations indicated supercomputer performance of approximately 200 million events per second using 1,536 processors in 2003. This number increased to 12.26 billion events per second on 65,536 processors in 2009, and 504 billion events per second in 2013 using almost 2 million cores. However, over this 10-year span, the performance *per core* increased by only a factor of two. Performance increases are being driven almost entirely by the exploitation of parallel processing.

Exploitation of massively parallel computer architectures presents many critical challenges to the modeling and simulation community. Perhaps the most obvious is the fact that the simulation computation must be developed in a way to exploit finer grains of computation, i.e., the atomic unit of computation that cannot be subdivided into computations that are mapped to different cores must become smaller. For example, in numerical simulations involving large matrix computations, rather than mapping entire rows, columns, or submatrices to individual cores, new approaches that consider mapping individual elements of the matrix to different cores are beginning to show promise, thereby exposing much higher levels of parallelism in the computation. The simulation computations and associated algorithms must be rethought to consider such fine-grained parallelism.

Once the simulation has been formulated as a fine-grained parallel computation, a key challenge concerns efficient execution of the simulator. Communication latency has long been a principle impediment to efficient execution of parallel simulations; keeping the numerous cores busy with useful computations becomes very challenging if the delay to transmit information between cores increases. The reason is because many computations will have to remain idle, waiting for results computed on other cores to arrive. This problem becomes even more challenging in fine-grained simulation computations where the amount of computation between communication actions becomes small. Latency hiding techniques that can mask communication delays become even more critical in order to successfully exploit large-scale parallel computers. Further, effective exploitation of memory system architectures becomes increasingly more important. When the state size encompassed by the simulator increases, efficient use of cache memory systems becomes increasingly more challenging and important.

Another key question concerns how to map the parallel simulation model to the parallel architecture, especially for simulations that are modeling highly irregular physical systems. For example, consider a simulation of a large network, such as the Internet. Many networks such as these that arise in natural and engineered systems are highly irregular and often contain "hub nodes" with high interconnectivity relative to other nodes in the network. The amount of activity and thus simulation computation can vary by several orders of magnitude from one network node to another. Partitioning and mapping large-scale irregular network simulations to execute efficiently on modern supercomputers is a challenging task that requires further exploration.

4.1.2 Parallel Simulation on Heterogeneous Computing Platforms

Modern computers ranging from supercomputers to mobile devices are increasingly being composed of combinations of general purpose processors coupled with hardware accelerators, e.g., GPUs. GPUs are hardware accelerators that implement specific computational tasks that would otherwise be performed by the central processing unit (CPU). They derive their name from the fact that they were initially developed to render graphics for display devices on workstations and personal computers. GPUs have since found broader application as a means to accelerate data-intensive numerical computations. High-volume manufacturing of GPUs has driven their hardware cost down, making them attractive components for high-performance computing systems.

GPUs are designed for data-parallel computations, i.e., computations where the same operations are applied to large volumes of similarly typed data. Computing elements are organized to implement single-instruction-stream and multiple-data-stream (SIMD) operations; i.e., a common program is executed by the many computing elements (cores) but operating on different data. This data-parallel processing is the main source of performance improvement in these hardware platforms.

A significant body of work has emerged in developing methods to exploit GPU architectures for numerical simulation applications. Such applications are often formulated as matrix computations, making them well suited for the exploitation of these architectures. Other computations such as discrete event simulations are typically not structured as matrix computations. Rather, they often utilize much more irregular data structures which are more challenging to map to GPU accelerators. It is especially significant that the next generations (at least) of the highest-end supercomputers will be designed as clusters of nodes, each of which is composed of a small number of multicore processors and advanced GPUs that share memory. The GPUs will offer the vast majority of the computational parallelism in these platforms.

To illustrate some of the challenges associated with executing irregular simulations on GPUs, consider a large discrete event simulation program that consists of many event computations. To execute efficiently in the SIMD style used in GPU architectures, the simulation should consist of relatively few, and ideally only one, type of event, a restriction that applies to a limited number of discrete event simulation applications. SIMD code is also most efficient when it is mostly straight-line code with few branches and when loops running in parallel take (almost) the same number of iterations. This is a problem for discrete event simulations, whose control flow frequently branches. Also, code running on GPUs, for at least the next few generations of them, will not be able to execute operating system code, perform I/O, or send or receive messages without involving the CPU. This last restriction is also a major problem for discrete event simulation since, and on average, each event has to send one event message. This will almost certainly cause the CPU to be a performance bottleneck in any discrete event simulation that attempts to run events in parallel with the GPUs. The traditional multicore/multiprocessor architectures will likely remain more efficient for these simulations, until such time as there is more convergence between CPU and GPU architectures than is currently specified in the technology road maps. Making efficient use of GPUs for highly irregular, asynchronous parallel, discrete event simulations will be, to say the least, a major challenge.

Concurrent execution can be obtained by partitioning the state variables of the simulation into objects and processing the same event computation concurrently across these objects. Further, as mentioned earlier, each event computation should contain few data-dependent branch instructions because if the program is executed over different data, the execution of different program sequences resulting from different branch outcomes must be serialized in the SIMD style of execution. The future event list used in discrete event simulations, a priority queue data structure, is similarly challenging to distribute for concurrent access on existing GPU architectures. Thus, restructuring irregular simulations for execution on GPUs remains challenging.

Further, once the computation has been reformulated for execution on a GPU, other computational challenges must be addressed. Specifically, the memory available within the accelerator remains limited, and moving data in and out of the GPU's memory can quickly become a bottleneck. Techniques to hide the latency associated with data transfers are essential to achieve efficient execution. Memory systems are typically organized as banks of memory, with concurrent access to memory distributed across different banks. However, accessing the data residing in the same bank must be serialized. Care must be taken to map the simulation's state and other variables to the memory system to avoid creating bottlenecks.

Effective exploitation of GPU architectures today requires careful programming that is tailored to the specific target architecture. This makes codes relatively brittle—performance optimizations designed for one architecture may no longer be valid when the next-generation architectures appear. Tools to automate the mapping of simulation computations to GPU architectures are needed to alleviate this task from the programmer. Moreover, software development on GPU architectures can be burdensome. Application-specific languages or advances in parallel compilers may offer ways to simplify the programming task.

4.1.3   Array Processors as a Platform for Modeling and Simulation

Advances in field programmable gate array (FPGA) technology have improved their speed, performance, and connectivity with other devices while lowering their power consumption, leading to their emergence as array processors for use in high-performance parallel computing platforms. These processors combine the features of application-specific integrated circuits with dynamic reconfigurability, especially during runtime, to provide a suitable system for performing massively parallel operations. These systems have a sufficient number of processing units to provide large-scale parallelism, higher processing power, and shorter reconfigurability time, even during the execution of the same program. Their performance is better than microprocessors by a factor of 100 and more (Tsoi and Luk 2011).

The suitability of array processors as a parallel processing platform has been and is being investigated in data-intensive applications, such as signal and image processing, database query, big data analysis, and applications that are compute and memory-intensive, such as high-speed network processing, large-scale pattern matching, influence-driven models for time series prediction from partial observation, model-based assessment, and many more (Dollas 2014).

In array processor architectures, a single-instruction controls the simultaneous execution of data in the processing units (SIMD, discussed earlier) which is efficient when data sets in the processing units do not rely on each other. The topology of the array processor is heavily influenced by the structure of the interconnection network, its speed of connectivity, and its configurability for a specific application. Because of this dependency, efficient partitioning (mapping) algorithms are needed.

Array processors have good potential for big data processing models, and good results have already been shown for some unique applications. However, their suitability for general applications and for modeling and simulation of complex systems needs to be studied. One of the reasons for the limitation of array processors is the coupling of prefetch instructions with the execution unit. Decoupling, along with the development of methodologies to properly map data dependency, needs to be researched.

Development of user-friendly programs for mapping models into the array architectures, creation of tools for dynamic reconfiguration of general application models, development of efficient dynamic routing algorithms to accelerate the routing phase specific for array processors, and production of an open-source hardware design to enable research into novel reconfigurable architectures are all very important. Selecting an appropriate memory model, such as shared, distributed, or hybrid, to develop an efficient programming interface for the selected memory models needs to be researched, especially for complex compute bound simulation models.

Developing hardware solutions for data processing that support high degrees of parallelism is challenging because as core counts increase, the average on-chip distance between arbitrary communication points also grows. Thus, enforcing scalable communication patterns is crucial. For example, an algorithm can be parallelized by replicating a task over many processing elements and organizing them into a feed-forward pipeline.

The creation of more high-level development environments, such as OpenCL in place of low-level ones (VHDL, Verilog), for high-level abstractions will allow array processors to be developed more efficiently independent of the technical advances of modern synthesizers and provide fundamental trade-offs between speed and chip space. Furthermore, appropriate trade-off analyses between speed and generality, clock speed and power consumption, chip area and accuracy, expressiveness, and (runtime) flexibility need to be made. In addition, while OpenCL provides programming portability, it does not provide performance portability. Therefore, the portability issue also needs to be resolved. Thinking outside the box while researching porting data processing algorithms from software to hardware and accurately abstracting the underlying operations of a given task, including synchronization, to achieve high degrees of parallelism and flexibility is also important (Woods 2014).

Other challenges include speeding up the tools for mapping a model description (which is in the range of hours to days) using existing parallel programming languages, such as OpenCL, CUDA (compute unified device architecture, a programming model to increase the computing performance-restricted for a certain hardware), and SystemC.

4.1.4 Modeling and Simulation in the Cloud

Cloud computing offers a means to make modeling and simulation tools much more broadly accessible than was possible previously. Cloud computing provides the ability to offer modeling and simulation tools as a service that can be readily accessed by anyone with an Internet connection. In principle, users of such tools need not own their own computers and storage to complete the simulations. This feature can be especially beneficial for simulation computations requiring high-performance computing facilities because the cloud eliminates the need for simulation users to manage and maintain specialized computing equipment, a serious impediment limiting widespread adoption in the past. The "pay-as-you-go" economic model for the cloud is attractive when computational needs are heavy during certain periods of time, but much less during others. However, there are certain challenges that must be overcome for the modeling and simulation community to maximally exploit cloud computing capabilities.

Virtualization technology is used extensively in cloud computing environments. Virtualization enables one to create a "private" computational environment where resources such as CPU, memory, and operating system services appear to be readily available to applications as virtualized components. Virtualization provides isolation between applications, thereby enabling physical computing facilities to be shared among many users without concern for programs interfering with each other.

Cloud computing presents certain technical challenges, especially for parallel and distributed simulations. A significant issue that has impeded greater exploitation of public cloud computing services concerns communication delay. Both latency and latency variance, i.e., jitter, may be high in cloud computing environments and significantly degrade performance. This problem could be alleviated by improved support from cloud providers for high-performance computing. Another approach is to design parallel and distributed simulations with better ability to tolerate latency and jitter in the underlying communication infrastructure.

A second issue concerns contention for shared resources in cloud computing environments, as users are typically not guaranteed exclusive access to the computing resources used by their programs. This can lead to inefficient execution of parallel and distributed simulation codes. An approach to addressing this problem is to develop mechanisms to make these codes more resilient to changes in the underlying computing environment during the execution of the simulation. For example, dynamic load adaption is one method that can be applied to address this issue.

Cloud computing introduces issues concerning privacy and security. These are issues that are well known in the general computing community and are equally important if cloud computing is to be successfully exploited by the modeling and simulation community.

There is a trend to recognize that groups of software services require different facilities and support from cloud computing, virtualization, and service-oriented architectures. Arguably this is also true in this area and is emerging as "Modelling and Simulation as a Service" (MSaaS). This could cover modeling and simulation applications ranging from "online" simulation, where multiple users can access the same simulation (and potentially share information among them), to simulations requiring various high-performance computing support, to groups of interoperable simulations to pipelines of simulations and supporting services (real-time data collection, simulation analytics, optimizers, etc.). These in turn make novel demands of cloud- and service-oriented architecture concepts such as workflow, orchestration, choreography, etc.

4.1.5 Mobile Computing Platforms

As discussed in the next section, the number of mobile computing devices dwarfs the number of desktop and server machines, the traditional platform used for modeling and simulation codes, and this gap is rapidly increasing. The increasing computing power of mobile devices means simulation codes need not be limited to remote servers or execution in the cloud. Rather, simulations can be embedded within a physical system itself. Increased use of mobile platforms such as drones provides many new opportunities for the use of simulation to monitor and assist in managing operational systems in real time. For example, simulations operating in drones may be used to predict the spread of forest fires or toxic chemical plumes, enabling one to dynamically adapt the monitoring process or institute approaches to mitigate damage. Transportation represents another important application where simulation embedded within the traffic network may be used to project traffic congestion arising after an incident and to explore alternate courses of action.

Embedding the simulation computations within the physical system being monitored or managed offers the opportunity for the computations to be used in tighter control loops using disaggregated data compared to approaches using back-end servers or the cloud. Further, placing the computations nearer to data streams lessens reliance on long-range communication capabilities and can mitigate privacy concerns by eliminating the need to communicate and store sensitive data on centralized servers.

Data-driven online simulations are growing in importance. Sensor data and analytics software process live data streams to construct or infer the current state of the system. Simulations are then used to project future system states, e.g., to improve the monitoring systems to better track the physical system as it evolves, or to be used as a means to optimize or improve the system. These simulations must run much faster than real time to be useful. Paradigms such as DDDAS can be expected to grow in use into the foreseeable future.

Mobile computing platforms present new challenges for simulation applications. The simulations must be able to produce actionable results in real time. This necessitates automation of many of the steps in a modeling simulation study. For example, input data must be analyzed and processed rapidly to parameterize and drive the simulation models. Experimentation plans for analyzing possible future outcomes must be rapidly created and executed. Simulation runs must be mapped to available computing resources, and analyses of output data must be completed and interpreted with minimal delay, and translated into action plans. Data from the physical system offers the opportunity to automatically calibrate, adapt, and validate simulations by comparing observed system behaviors with those predicted by the simulations.

Energy consumption is another area of increasing concern. In mobile computing platforms, reducing the energy required for the computation will increase battery life and/or can allow smaller, more compact batteries to be used. However, most of the work to date in energy consumption has focused on low-level hardware, compiler, and operating system issues. Relatively little work has been completed to understand the energy consumed by simulations. Better fundamental understandings of the relationship among energy consumption, execution time, data communications, and model accuracy are needed. These relationships must be better understood for both sequential and parallel/distributed simulations. Approaches to optimize energy consumption consistent with the goals and constraints of the simulations in terms of timeliness in improving results are needed.

4.1.6 Neuromorphic Architectures

Neuromorphic computers are a radical departure from the traditional von Neumann computer architectures. They are essentially the hardware realization of neural nets, modeled loosely on animal nervous systems, and are capable of doing tasks such as image processing, visual perception, pattern recognition, and deep learning vastly more quickly and energy efficiently than is possible with traditional hardware.

At this point, it is too early to do more than speculate about how neuromorphic computation will be incorporated into simulations once they become better understood and more widely available. But one possible use case might be in autonomous vehicles such as self-driving cars or aerial drones which may need extensive image processing in embedded simulations to predict in real time the behavior of other nearby vehicles. Those tasks may not be directly programmed in a traditional rule-based manner, but may instead make use of the learning capability inherent in neuromorphic chips to adapt to local conditions and to the world as it changes over time. Another example is a simulation of a system where visual processing is critical, such as satellite aerial surveillance. The systems themselves may make use of neuromorphic computation, but *simulations* of those systems will likely need it as well, since otherwise the simulation will probably run many times slower.

4.2 Pervasive Simulation

4.2.1 Embedding Simulations (into literally everything)

A key observation in the "post-digital revolution society" is that information and communication technologies (ICT) have become *pervasive*, i.e., interwoven with human behavior, or in other words: the "fabric of everyday life" to such an extent, that the separating view of a "physical world" being connected with a "digital world" is ceasing. Today, we talk about one "cyber-physical" world (Cyber-Physical Systems, an NSF program developed by Helen Gill in 2006), referring to the tight entanglement of real-world physical objects (things, appliances) and processes (services), with their digital data representation and computations in communication networks (the "cyber"). Embedded, wirelessly connected tiny compute platforms equipped with a multitude of miniaturized sensors collect data about phenomena, analyze and interpret that data in real time, reason about the recognized context, make decisions, and influence or control their environment via a multitude of actuators. Sensing, reasoning, and control, thus, are tightly interconnecting the physical and digital domains of the world, with feedback loops coupling one domain to the other. Connecting the "physical" with the "digital" based on embedded electronic systems, which in addition to executing preprogrammed behavior also execute simulations we call *pervasive simulation*. Pervasive simulations have clear synergies and overlaps with mobile and dynamic data-driven application systems discussed earlier.

4.2.2 Collective Adaptive Simulations

Taking the plenty-hood of todays embedded platforms with their computational, sensory, reasoning, learning, actuation, and wireless communication capacities (smartphones, autonomous vehicles, digital signage networks, stock exchange broker bots, wearable computers, etc.), it is not just considered possible, but already a reality that these are programmed to operate cooperatively as planet scale ensembles of collective adaptive computing system (CAS). CAS research asks questions on the potential and opportunities of turning massively deployed computing systems into a globe-spanning "superorganism," i.e., compute ensembles exhibiting properties of living organisms such as displaying a "collective intelligence." Essential aspects of CAS are that they often exhibit properties typically observed in complex systems, such as (i) spontaneous, dynamic network configuration, with (ii) individual nodes acting in parallel, (iii) constantly acting and reacting to other agents, and (iv) highly dispersed and decentralized control. If there is to be any coherent behavior in the system, it must emerge from competition and cooperation among the individual nodes, so that the overall behavior of the system is the result of a huge number of decisions made every moment by many individual entities. Pervasive simulation raises CAS to *collective adaptive simulations*.

4.2.3 Massive Scale Pervasive Simulations

The International Telecommunication Union (ITU) predicts there will be 25 billion devices online within the next decade, outnumbering connected people 6-to-1 (International Telecommunication Union 2012b). This will lead to a pervasive presence around us of objects and things (e.g., RFID tags, sensors, actuators, mobile phones), which will have some ability to communicate and cooperate to achieve common goals. This paradigm of objects and things ubiquitously surrounding us is called the Internet of Things (IoT). The ITU defines IoT as a "global infrastructure for the information society, enabling advanced services by interconnecting (physical and virtual) things based on, existing and evolving, interoperable information and communication technologies" (International Telecommunication Union 2012a). The IoT covers different modes of communication, including between people and things, and between things (machine-to-machine or M2 M). The former assumes human intervention and the latter none (or very limited).

4.2.4 Privacy, Security and Trust

A primary aim of IoT is to deliver personalized or even autonomic services by collecting information from and offering control over devices that are embedded in our everyday lives. The reliance of IoT on simple, cheap, networked processors has implications for security; the potentially invasive nature of the information gathered has implications for privacy; and our reliance on machine-to-machine systems to make decisions on our behalf makes mechanisms for expressing and reasoning about trust essential. The need for trust has long been recognized, as stated recently by Moulds (2014), the "… pivotal role in … decision making means it is essential that we are able to trust what these devices are saying and control what they do. We need to be sure that we are talking to the right thing, that it is operating correctly, that we can believe the things it tells us, that it will do what we tell it to, and that no-one else can interfere along the way."

As pervasive simulations become more commonplace, it is essential that they be secure or at least tolerant of cyber threats. Privacy and trust issues must be adequately addressed to realize widespread adoption.

As the world becomes more connected, we will become dependent on machines and simulations to make decisions on our behalf. When simulations use data from sensors, devices, and machines (i.e., things) in the network to make decisions, they need to learn how to trust that data as well as the things with which they are interacting. Trust is the belief in the competence of a machine or sensor to act dependably, securely, and reliably within a specified context (Grandison and Sloman 2000). Trust is a broader notion than information security; it includes subjective criteria and experience. Currently securing sensors and devices is accomplished through information security technologies, including cryptography, digital signatures, and electronic certificates. This approach establishes and evaluates a trust chain between devices, but it does not tell us anything about the quality of the information being exchanged over time.

As the number of sensors connected to the network grows, we will see different patterns of communication and trust emerges. If we assume a hierarchical connection of components, sensors are at the end-nodes, which communicate data to aggregators. Sensors may be unintelligent (sense environment and send data to aggregator) or they may be intelligent (sense environment, reason about the data, and communicate with aggregators). Aggregators are capable of collecting data from sensors, reasoning about that data, and communicating with other aggregators. Having aggregators communicate with each other enables trust decisions to be made in a more distributed manner, reasoning about trust across geographic areas.

Data from the sensors and aggregators will be fed into models and simulations that are making predictions and/or decisions that will impact our lives. Data from sensors or aggregators may be in conflict with each due to malfunction, bad actors, tampering, environmental conditions, context conditions, and so on. Whether or not the simulation should trust this data must be established by an agent that is capable of a trust evaluation prior to it being deemed useful as information. Further, if simulations have a role in controlling or giving commands to some sensor, actuator, or device in the IoT system (i.e., a cyber-physical system), then the data the simulation uses from external sources in which to make those decisions must be trustworthy such that it is not purposely misled into issuing malicious commands.

4.2.5 Foundational Research Concerns

In order to develop a deep scientific understanding of the *foundational principles of pervasive simulations*, we need to understand the trade-offs between the potentials of top-down (by design) adaptation means and bottom-up (by emergence) ones, and possibly contributing to smoothing the tension between the two approaches. We need to understand how and to what extent pervasive simulations—when involving billions of components—can create a "power-of-masses-principle" and possibly express forms of intelligence superior than that of traditional artificial intelligence. Furthermore, understanding properties concerning the evolutionary nature of pervasive simulations, e.g., open-ended (unbounded) evolutionary simulation systems, the trade-off and interaction between learning and evolution, and the effect of evolution on operating and design principles are of foundational importance. Understanding the issue of pluralism and diversity increase in *complex pervasive simulation systems* as a foundational principle of self-organization, self-regulation, resilience, and collective intelligence is needed. Last but not least, laying down new foundations for novel complex *pervasive simulation theories* for complex, adaptive, large-scale *simulation superorganisms* (including lessons learned from applied psychology, sociology, and social anthropology, other than from systemic biology, ecology, and complexity science) remains a key challenge for the scientific community.

4.2.6 Systems Research Concerns

In order to develop principles and methods for the design, implementation, and operation of *globe-spanning simulation superorganisms,* we identify systems research concerns such as (i) Opportunistic Information Collection: Systems need to be able to function in complex, dynamic environments where they have to deal with unpredictable changes in available infrastructures and learn to cooperate with other systems and human beings in complex self-organized ensembles. (ii) Living Earth Simulation: The provision of a decentralized planetary-scale simulation infrastructure strongly connected to the worlds' online-data sources (search engines, power grids, traffic flow networks, trade centers, digital market places, climate observatories, etc.) is needed as a means to enable a model-based scenario exploration in real time—at different degrees of detail, varying timescales, integrating heterogeneous data and models. (iii) Collaborative Reasoning and Emergent Effects in Very-Large-Scale Pervasive Simulations: Reasoning methods and system models are needed that combine machine learning methods with complexity theory to account for global emergent effects resulting from feedback loops between collaborative, interconnected simulations. (iv) Value-Sensitive Simulations: Research is needed on ethics, privacy, and trust models for simulations that are robust and resilient to common threat models in planetary-scale simulations.

Toward *pervasive simulation applications*, we have to look at the specifics of design, implementation, and operational principles rooted in the very nature of application domains of societal relevancy: e-health ecosystems, fleets of self-driving vehicles, reindustrialization (Industry 4.0), physical internet (intelligent logistics), digital economy, energy management and environmental care, citizen science, combinatorial innovation, liquid democracy, etc.

4.3 New and More Complex Simulation Paradigms: A Plethora of Models

As we simulate ever larger and more complex systems, from thousands of interacting components to millions and billions, the complexity of the models we must build and execute also dramatically increases at all levels in the layered stack of simulation software. Large simulations must frequently combine different modeling paradigms and frameworks, at different temporal and spatial scales, and different synchronization and load balancing requirements. They must interface with other non-simulation software, such as databases, analysis packages, visualization systems, and sometimes with external hardware systems or humans. And a single execution of a model is not sufficient. We always need a structured ensemble of many independent executions in a proper simulation study. Among the greatest R&D challenges we face is the creation of simulator platforms, frameworks, tool chains, and standards that allow this diversity of paradigms to interoperate in a single simulation application. The software challenges are made even greater because of the hardware computing architectures on which complex simulations run are also changing rapidly, as was discussed earlier.

4.3.1 Complex Simulations

Simulations are becoming ever more complex as they are applied to new domains and grow in scale and fidelity required. The additional complexity is multidimensional, often with multiple kinds of complexity in the same model, leading to a variety of architectural requirements beyond correctness, fidelity, and performance. Examples of this complexity include the following:

- *Federated models*—models composed (recursively) of separately developed submodels that are then coupled in structures that mimic the way the real systems being simulated are composed of coupled subsystems.
- *Multiparadigm models*—models that contain subsystems designed according to different paradigms, e.g., queuing models coupled to Petri net models and numerical differential equation models.
- *Multiscale models*—models with significant phenomena occurring at different time or space scales, often differing by orders of magnitude.
- *Multiphysicsmodels*—models in which multiple different physical phenomena, e.g., fluids, solids, particles, radiation, fields, all coexist and interact.
- *Multiresolution models*—models in which it is necessary to be able to adjust a resolution parameter to allow a trade-off between time and space resolution or degree of detail for improved performance.
- *Multisynchronization models*—parallel models that use multiple synchronization paradigms, e.g., a hybrid or federation of different time-stepped, conservative event-driven, and/or optimistic event-driven synchronization algorithms.
- *Mixed discrete and continuous models*—models in which some parts are described by numerical equations describing state changes that are continuous in time, while other parts are discrete, in which all state changes are discontinuous in time.
- *Real-time models*—models that must produce results by specific real-time deadlines, often in embedded systems.
- *Hardware-in-the-loop (HWIL)*—a special case of a real-time model in which a simulation is coupled to a physical device with which it must synchronize and communicate at a real-time speed determined by the needs of the device.
- *Human-in-the-loop*—a simulation that interacts with humans in real time, at speeds and with response times keyed to human behavior and reaction times, and with I/O keyed to human sensory and action modes.
- *Models as components of other computations*—models that are subsystems of a larger computation that is not itself a simulation, e.g., an animation system, or a control system.

- *Models containing large non-simulation components*—for example models that run whole (parallel) speech understanding or visual systems inside a single event.
- *Virtual machines as model components*—an important special case in which the system being simulated involves computers or controllers that run particular software, and the execution of that software has to be faithfully duplicated, including timing, for the simulation to be correct.

Some of these complexity dimensions are reasonably well understood, at least in principle. But others are poorly understood even theoretically, and considerable research will be required to clarify them. In most of these dimensions, there are no widely accepted standards and no *robust tool chains* for building, executing, debugging, or validating them. When such models are built today, they are usually one-offs that are dependent on the specific details of the particular application needs and are likely to contain ad hoc design decisions and engineering compromises that make the simulation brittle, unportable, and/or unscalable.

To manage this kind of simulation complexity, we need to develop simulation-specific software engineering standards, abstractions, principles, and tools. For example, one step might be to define *standards* for simulation software as a stack of software layers, in which each layer provides and exposes additional services for use by the layers above, and abstracts or hides some features from the layers below. This follows the pattern set by the layering of general purpose system

Table 4.1 Abstraction layers of a simulation software stack

Abstraction layer	Function
Model layer	The code of a particular model (or component)
Model framework layer	Collection of model classes, components, and libraries for a single application area, such as network simulation or PDE solution
Simulator layer	Provides a single paradigm for simulation time, space, naming, parallelism, and synchronization for use in one component of a (possibly) federated simulation
Component federation layer	Provides interface code to allow independently created submodels, possibly written in different languages, to communicate, synchronize, and interoperate in various ways to become a single federated model
Load management layer	Within one parallel model execution, measures resource utilization (time, energy, bandwidth, memory) at runtime and dynamically manages or migrates loads to optimize some performance metric
Ensemble layer	Runs many instances of the same model as an ensemble in a single large job, for such purposes as parameter sensitivity studies, parameter optimization, variance estimation. Handles scheduling, failures, accounting, and time estimates, allocates file directories, decides on ensemble termination, etc.
Operating system /job scheduler layer	Runs independent jobs in parallel. Provides processes, interprocess communication, I/O, files systems, etc.

and application software, or the TCP/IP and OSI protocol stacks. Table 4.1 shows a set of abstraction layers that might crudely exemplify such a simulation software stack.

Presumably there could be various alternative systems at each layer, just as there are different protocols at each layer of the TCP/IP stack. The point is not to suggest this particular organization for the simulation stack. Any such standard should be the result of lengthy and careful consideration among the stakeholders in the simulation community, perhaps under the auspices of professional societies such as the Association for Computing Machinery (ACM). But there is an urgent need for software engineering principles specific to simulation to help manage the complexity that currently limits the kinds of simulations we can realistically attempt.

4.3.2 Unification of Continuous and Discrete Simulation

One of the basic simulation questions that will require considerable research to clarify is the relationship between continuous and discrete simulations. On the surface, they seem strikingly different. Continuous simulations treat state changes as *continuous* in time, whereas discrete simulations treat state changes as *discontinuous*. The fidelity of continuous simulation is dominated by numerical considerations (error, stability, conservation, etc.), whereas for discrete models it is dominated by detailed correspondence with the system being modeled and also by statistical considerations. The two kinds of simulation are sufficiently different that there are very few people who are expert in both.

Despite the differences, it is common for complex models to combine aspects of both discrete and continuous submodels. Frequently, for example, one part of a system, e.g., electric power distribution, or aircraft aerodynamics, is described by differential equations and represented by a continuous simulation, but the digital control system for those same systems (power grid, aircraft) is better represented by discrete models. The entire coupled simulation is thus a mix of continuous and discrete models.

The problem in coupling continuous models to discrete ones is that the continuous side is usually programmed as a time-stepped simulation, while the discrete side is likely to be event-driven, and the two do not share a common synchronization mechanism. To start with, we need robust, high-performance parallel integration algorithms on the continuous side of the coupling that can freely accept inputs from the event-driven discrete side at arbitrary (unpredictable) moments in simulation time falling between two time steps. Some, but not all, integration algorithms have the property that at any simulation time one or more new, shorter time steps can be interpolated between two preplanned ones without loss of accuracy or other key properties. However, in practice, even if the integrator has that property, the actual integration codes were not developed with that capability in mind and they do not incorporate the necessary interpolators and synchronization flexibility.

A more ambitious research agenda is to *unify* the theory and practice of parallel continuous and parallel discrete event simulation. A few dozen scattered papers have been published related to this theme, but the issue is still not widely recognized and it will certainly require a major international research and development effort to clarify the issues and build appropriate tools. Unification would require development of a variety of scalable parallel variable rate integrators, both explicit and implicit, that are numerically stable. They need to support variable, dynamically changing spatial resolution as well (in the case of PDE solvers). To execute optimistically, or couple efficiently to an optimistically synchronized model, a unified discrete and continuous simulator will have to support rollback as well.

4.3.3 Co-simulation and Virtual Simulation

Co-simulation stems from modeling embedded systems where verification of hardware and software functionality as a system is performed simultaneously before and during the design phase to ensure the final manufactured system will work correctly. The approaches developed for co-simulation, as well as the tools developed for describing the models and simulating the correct functionality, use VHDL, VERILOG, and SystemC. The use of these tools and methodologies for co-simulation of embedded systems is rather mature. Although some of the existing tools such as VHDL and Verilog do not have standard interface features for communication between hardware and software, SystemsC and some industry-developed tools used in the design of this type of system have been in practice for some time.

Compared with the digital realm, hybrid modeling and co-simulation of continuous and discrete systems and their synchronization have not been addressed to the level necessary to provide accurate results when used for the design of mixed and hybrid complex systems. The complexity of continuous/discrete systems makes their co-simulation and validation a demanding task and the design of heterogeneous systems challenging. The validation of these systems requires new techniques offering high abstraction levels and accurate simulation from a synchronization and intersystem communication point of view. This is especially necessary for the development of cyber-physical systems, which are a combination of continuous components that may be defined by a set of ordinary or partial differential equations, discrete components (such as microcontrollers) for control purposes, and embedded software for local and remote operation via the Internet.

Appropriately sharing design parameters between discrete and continuous subsystems for access by either subsystem at the correct time, occurrence of the correct event, and initiation of events by either subsystem are among the issues needing to be researched. In addition, the following issues require further research: scheduling events to occur at a specific time (time events) or in response to a change in a model (state events); events that are described with predicates (Boolean expressions), where the changing of the local value of the predicate during a co-simulation triggers the event; modeling abnormal behavior, such as those caused by a random event, such as faults or misuse; and defenses against the misuses including fault tolerance mechanisms for protection against them.

One of the most important difficulties in continuous/discrete co-simulation is the time synchronization between the event-driven discrete simulation and the numerical integration into the continuous solver which influences the accuracy and the speed of the simulation. The exchange of events between the discrete and the continuous models is especially critical for co-simulation. The continuous model may send a *state event*, whose time stamp depends on its state variables (e.g., a zero-crossing event), and the discrete model may send events, such as *signal update events*, that may be caused by the change of its output or the *sampling* events (Nicolescu et al. 2007 and Gheorghe et al. 2007).

Since designs are becoming more complex, it is expected that some of the methodologies developed for co-simulation of embedded systems will be adaptable for co-simulation of cyber-physical systems and will verify the stability, controllability, and observability of such systems under various operating environments. Challenges in this area particularly include the development of a single tool for co-simulation of continuous, discrete, and embedded software components of cyber-physical systems to ease synchronization events needed among these three subsystems.

4.3.4 Simulation Ensembles

A complex model code will almost always have inputs that describe initial conditions, parameters that control or modulate the system's behavior during simulation time, and random seeds that initialize random variables controlling stochastic behavior. A serious simulation study involves hundreds, thousands, or even more executions of the same model code, i.e., an *ensemble* of simulations, to explore and quantify the behavioral variation that the model can produce.

Ensembles of simulations are required in at least these circumstances, and others as well:

- to broadly explore and survey that space of behaviors that the model produces with different inputs and parameters;
- to examine the sensitivity of model behavior to perturbations of inputs or parameters;
- to find optimal input parameter values that maximize some output metric;
- to measure the mean, variance, correlation, and other statistical properties of the models' outputs over a large number of executions with different random seeds;
- to search for or measure the frequency of rare events that can occur in the behavior of the model;
- to conduct uncertainty quantification studies;

- to guide and track training progress in human-in-the-loop training simulations;
- to do runtime performance studies, including scaling studies.

Because ensemble studies are almost universal, the simulation community should recognize *ensemble studies as the fundamental unit of simulation*, rather than concentrating primarily on the single execution. It is the resources and costs required of the ensemble that matters, not those required for any individual run. Thus, to reduce resource utilization it is often much more important to optimize the number of simulations in the ensemble rather than the performance of individual simulations. Likewise, if time to completion of the entire study is critical, then it is more important to design the study so that more parallelism is derived from running simulations in parallel, even if that means reduced (or no) parallelism within a single simulation.

Methods to automate the creation and execution of simulation experiments from job submission to resource allocation to execution of computational experiments including multiple runs are needed. We need to define standards and build tools to support the ensemble level of simulation. We should be able to run a *single job* in a form portable to multiple platforms, to conduct an entire ensemble study, or at least a good part of it. It should choose inputs, parameter values, and random seeds, dynamically allocate nodes on a parallel machine as needed, launch individual simulations on those nodes, calculate their time estimates, allocate file system space for their outputs, monitor their normal or abnormal termination, and decide as some simulations terminate and free nodes, which next simulations to run. The code or script that manages the ensemble may be interactive or may conduct the entire sensitivity or optimization or variance estimation study autonomously, deciding what simulations to run, in what order, and when to stop.

4.4 Beyond Big Data: Synergies Between Modeling, Simulation and Data Analytics

Data analytics and machine learning algorithms provide predictive capabilities, but are fundamentally limited because they lack specifications of systems behavior. Modeling and simulations fill this gap, but do not exploit new capabilities offered by new machine learning algorithms. Approaches that synergistically combine these methods offer new approaches for system analysis and optimization.

4.4.1 Simulation Analytics

Traditional simulations compute aggregated statistics that are reported back to analysts. Recently, new HPC-based approaches for analysis of disaggregated data and sample paths through simulations exist, potentially providing much finer grained analysis results. As an example, large-scale high-fidelity multiagent simulations are being increasingly used in epidemiology, disaster response, and urban planning for policy planning and response. These simulations have complex models of agents, environments, infrastructures, and interactions. The simulations are used to develop theories of how a system works, or carry out counterfactual experiments that entail the role of various "system-level interventions." In this sense, simulations can be thought of as theorem provers. They are also used for situation assessment and forecasting. The eventual goal in each case is to design, analyze, and critique policies. The policies can be viewed as decisions taken by policy makers ranging from small groups to local and national government agencies. To use simulations in this setting, one often resorts to carrying out statistical experiments; these experiments can be factorial style experiments, but they can also be sequential experiments and use adaptive designs. Computation trees are examples of this. Even a moderate-sized design leads to a large number of runs; this coupled with simulations of systems at scale produce massive amounts of data.

As simulations become larger and more complex, however, we encounter a number of challenges. First, if a simulation is too computationally intensive to run a sufficient number of times, we do not obtain the statistical power necessary to find significant differences between the cells in a statistical experiment design. Second, if the interventions are not actually known ahead of time, we do not even know how to create a statistical experiment. This can be the case, e.g., when the goal of doing the simulation is to find reasonable interventions for a hypothetical disaster scenario. Third, as the system evolves in time, it is often necessary to incorporate new information leading to interactive systems with certain real-time requirements. New methodologies and new techniques are needed for the analysis of such complex simulations. Part of the problem is that large-scale multiagent simulations can generate much more data in each simulation run than goes into the simulation, i.e., we end up with more data than we started with. Although we have discussed these issues in the context of multiagent simulations of large socially coupled systems, they are applicable to other classes of biological, physical, and informational simulations as well. Additional discussion can be found in Marathe et al. (2014) and Parikh et al. (2016).

A broad challenge is that of *sense-making*. The basic issue is actively studied in artificial intelligence (AI). As discussed above, there are three parts to the problem. *Simulation analytics pertains to developing new algorithmic and machine learning techniques that can be used to support the above tasks.* They involve the following: (i) how to design a simulation that computes the right thing and summarization of simulation data, (ii) finding interesting patterns in the data sets, (iii) discovering potentially new phenomenon—how to analyze simulation results to extract insights, and (iv) integrating the data with real-world observations to provide a consistent partial representation of the real world under study. We discuss each of these topics below; also see (Barrett et al. 2011, 2015).

- Summarization: Summarization of simulation data is needed as massive amounts of data are being produced by these systems; it is expected that the simulated data can be orders of magnitude larger than the data that was used to drive the simulations. What does summarization mean in this context? How does one summarize the data? How does one retain important information and find it in the first place? The challenges here lie in developing and adapting statistical science and machine learning techniques on the one hand and algorithmic techniques on the other. The basic topic of summarization has now been studied in the data mining literature. Summarizing simulation-based data can of course use these techniques but also has aspects that might facilitate the development of specific methods.

- Finding Interesting Patterns: An important question arising within the context of simulation analytics is to identify interesting patterns. These patterns might point to anomalies or help with summarization or help discover potentially new phenomenon. The question here revolves around data representations and pattern representations, and efficient and provable algorithmic techniques to find these patterns.

- Discovering Potentially New Phenomenon: This is related to the previous problem but takes into account the problem semantics to discover a potentially new phenomenon. For example, a good pattern finding algorithm might be able to find clusters of a certain size repeated in certain simulations. Knowing that these simulations pertain to epidemiological outbreaks or star formation might provide new clues on super spreaders in a social network.

- Information Synthesis: An overarching problem is to try and synthesize the simulated information produced by different simulation components. The synthesis of data is an important issue, and one could view simulations as an approach to build a coherent view of bits and pieces of data, i.e., gathered by measuring aspects of the real world (either systematically or as a part of convenience data). In this context, the notion of information needs to be broadened not just as numerical data but also procedural and declarative data, information that pertains to how things work or how things behave. Indeed, simulations provide a natural way to both interpolate sparse data to form a coherent view but also allow us to extrapolate this data to develop potential possible worlds. Information synthesis comes up in physical systems but is most apparent when dealing with modeling and reasoning about biological, social, and informational systems, e.g., urban transportation systems, public health, banking, and finance.

- Believability: Why should one believe simulation results? There is much discussion of this topic in other sections, and hence, the focus here is on developing methods that can allow policy analysts a way to see patterns produced by simulations that can increase confidence in the simulation results. For instance, stochastic simulations might produce many possible branches as the simulation evolves. Is there a way to summarize these branches so that we can make sense of why this might have happened?

4.4.2   Model and Data Representation

Integrating algorithms for simulation and data raises questions concerning the most effective representation of models and data. Representations are needed to facilitate use of formal methods. Another challenge concerns the creation of domain-specific languages and efficient translators to accelerate the creation and exploitation of models. Data and model representations have been studied in other contexts as well, including databases, digital libraries, semantic web. Many of the concepts studied there are equally applicable. We will focus on the following topics: (i) the physical and logical way to store and represent numerical, declarative, and procedural data; (ii) formal methods to reason about the data and models within the simulation environment; and (iii) data and model synthesis for coherent system representation. The digital library and semantic web community has made significant advances in this area. Informally digital libraries refer to systematic organization of data and associated data along with methods to coherently access these data sets. In this sense, digital libraries are different from traditional databases. They are usually built on top of a logical representation of data in the form structured or semi-structured data; see (Ledig et al. 2011a, b).

An important research direction is to develop digital library concepts and frameworks to support simulation and modeling. This would require the following: (i) logical and physical organization of data from raw data sets that may be distributed across different locations to structured (e.g., RDBMS) and semi-structured data sets that provide a logical organization of data using the Resource Description Framework (RDF) and its extension; (ii) a hierarchy of progressively rich services for content generation, curation, representation, and management; and (iii) languages and methods to describe and develop complex workflows for integrating raw and simulated data sets, and manipulating these data sets while keeping the efficiency of system in mind. See Fig. 4.1.

Logical methods based on traditional database concepts have been useful in this context. Jim Gray and his colleagues make an excellent argument to used databases to organize the input as well as output data (Hey et al. 2009). This context can be extended not just to support organization of the data but to actively guide simulation during execution; these database-driven simulations provide a new capability in terms of expressiveness and human efficiency without compromising overall system efficiency. The use of RDF and its extensions to store and manipulate data is very promising—indeed graph databases have become extremely popular for storing certain kinds of data sets. The trade-off among extensibility, expressiveness, and efficiency between these representations is a subject of ongoing research (beyond purely theoretical terms).

Services:

As discussed in Leidig et al. (2011a, b, c), and Hasan et al. (2014), minimal digital libraries (DL) are expected to provide a set of DL services that meet the anticipated use case scenarios. User groups will be comprised of domain scientists who will use these services to generate complex work flows to support policy designs. Metadata

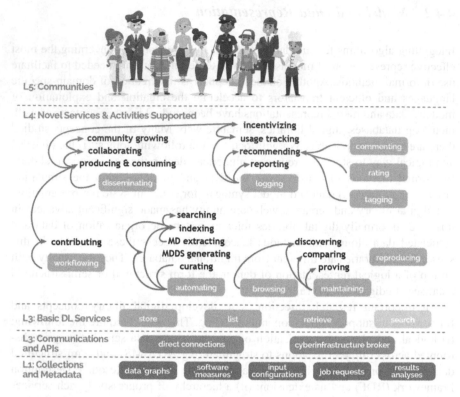

Fig. 4.1 Conceptual layers based on 5S framework for a digital library to support simulations

structures and provenance information connect input data and simulation data along with policy-related experimental metadata. A list of services is shown in Fig. 4.2. The idea is that these services form a rich and composable set of APIs that are in a sense organized to progressively support higher-level services within the list.

4.4.3 Large-Scale Data and Stream Management

The introduction of large-scale data and data streams within simulations present new data management challenges. Completing this work is a recent initiative on this topic (see streamingsystems.org/finalreport.html). Four basic topics were identified: (i) programming models, (ii) algorithms, (iii) steering and human-in-the-loop, and (iv) benchmarks. Much of the discussion applies to issues pertaining to the current document, and as a result, we will keep the discussion brief.

SIMULATION DIGITAL LIBRARY SCIENCE

Accessing	Acquiring	Annotating	Browsing	Cataloging
CI Connections	Classifying	Collaborating	Computation Reducing	Contextual Searching
Crawling	Customizing	Curating	Description Set Designing	Digitizing
Discovering	Educating	Experiment Supporting	Extracting	Evaluating
Federating	Filtering	Indexing	Information Lifecycling	Managing
Ontcology Building	Organizing	Preserving	Provisioning	Ranking
Rating	Recommending	Reproducing	Reusing	Searching
Simulation Connecting	Submitting	Tagging	Validating	Verifying

Fig. 4.2 Examples of digital library services needed to support complex simulations

As such, streaming in the context of M&S applies to streaming data that might be used to feed the simulations, e.g., data arriving for sensors as a part of the IoT vision that measure attributes of a social or a physical system. But streaming also applies to computing and reasoning about simulated data; often the size of these data sets is prohibitive. As a result, they are best viewed as streams and need to be processed on the fly to produce meaningful summarization.

4.4.4 Qualitative Modeling

Modeling physical systems and the decision-making process based on the results of simulations using those models are, for the most part, based on numerical quantities which quantitatively describe the relationship between inputs and outputs of the system. While such models are adequate for physical systems, they fail to meet the needs of the simulation community in regard to modeling and simulation of complex systems, such as those in cognitive science, knowledge engineering, health sciences, artificial intelligence, and many more. In these systems, models that describe the relationship between the inputs and observed outputs are in qualitative or linguistic form. This class of models is closer to human thinking than quantitative models and is easy to understand.

While the properties of these models are a better fit with human thinking and diagnosis in human terms, no efficient computational algorithms for the construction and execution of these models currently exist. Efficient tools to support the development of these models for qualitative data mining and feature extraction, pattern recognition, sensors, and sensor networks are in their infancy, especially when complex systems are considered.

Designing complex dynamic systems that require skills obtained through experience and the issues involved in translating human skills into the design of automatic controllers are challenges that remain. The transfer and reconstruction of skills may be performed using traces that were collected from an operator's actions (Bratko and Suc 2003). However, such transfer has only been tried for simple systems, and their suitability for complex systems needs to be tested.

Similar to skill transfer, the characterization of intuitive knowledge of the physical world, advanced methods for reasoning which use that knowledge to perform interesting and complex tasks, and development of computational models for human common sense reasoning need to mature through efficient and better tools.

Discretization is used to convert things which can be represented and reasoned about symbolically. It provides a means of abstraction to develop models for situations involving only partial knowledge where few if any details are known. In these cases, qualitative models can be used for inferring as much as possible from minimal information. For example, "We are at McDonalds" versus "We are at McDonalds at Manhattan Ave," (Forbus 2008). Challenges remain in developing models which can represent such partial knowledge in more descriptive form, and the tools for building models capable of understanding analogies and metaphors still need to be researched.

Some of the approaches to qualitative modeling include qualitative mathematics which are simple and can build the right model for a given situation (Bratko and Suc 2003). However, these models lack generality (i.e., each case requires a new model), and not all of the model-building skills are captured in the characterization.

The other challenges which remain in qualitative modeling are related to relevance, ambiguity, ontology modeling, and mature qualitative mathematics for complex system modeling. Ontology modeling based on traditional mathematics tends to be informal where informal decisions are used to decide what entities should be included in a situation, what phenomena are relevant, and what simplifications are sensible. Qualitative modeling will make such implicit knowledge explicit by providing formalisms that can be used for automating (either fully or partially depending on the task) the modeling process itself.

The ontological frameworks for organizing modeling knowledge, research on automatically assembling models for complex tasks from such knowledge, and application of qualitative models of particular interest to cognitive scientists, including how they can be used to capture the expertise of scientists and engineers and how they can be used in education, remain to be addressed. Open questions focusing on the relationships between ideas developed in the qualitative modeling and other areas of cognitive science need to be addressed.

Still challenges remain for accurate and precise mixed quantitative/qualitative modeling and simulation for applications in complex systems, such as those in health sciences (e.g., the cardiovascular system) and cognitive science, which require further basic and fundamental research (Nebot et al. 1998).

References

Barnes, P.D., C.D. Carothers, D.R. Jefferson, and J.M. LaPre. 2013. Warp speed: Executing time warp on 1,966,080 Cores. *Principles of Advanced Discrete Simulation* 327–336.

Barrett, C., S., Eubank, A. Marathe, M. Marathe, Z. Pan, and S. Swarup. 2011. Information integration to support policy informatics. *The Innovation Journal* 16 (1): 2.

Barrett, C., S., Eubank, A. Marathe, M. Marathe, and S. Swarup. 2015. Synthetic information environments for policy informatics: A distributed cognition perspective. E. Johnston (Ed.), *Governance in the Information Era: Theory and Practice of Policy Informatics.* New York: Routledge, 267–284.

Bratko, I., and D. Suc. 2003. Data mining. *Journal of Computing and Information Technology. CIT* 11 (3): 145–150.

Darema, F. 2004. Dynamic data driven applications systems: A new paradigm for application simulations and measurements. In *International Conference on Computational Science.*

Dollas, A. Big data processing with FPGA supercomputers: Opportunities and challenges. In *Proceedings of 2014 IEEE Computer Society Annual Symposium on VLSI,* 474–479.

Forbus, K.D. 2008. Qualitative modeling. In *Handbook of Knowledge Representation.* Chap. 9. Elsevier B.V.

Fujimoto, R.M. 2016. Research challenges in parallel and distributed simulation. *ACM Transactions on Modeling and Computer Simulation* 24 (4).

Gheorghe, L., F. Bouchhima, G. Nicolescu, and H. Boucheneb. A formalization of global simulation models for continuous/discrete systems. In *Proceedings of 2007 Summer Computer Simulation Conference,* 559–566. ISBN:1-565555-316-0.

Grandison, T. and M. Sloman. 2000. A survey of trust in internet applications. *IEEE Communications Surveys and Tutorials* 3 (4): 2–16.

Hasan, S., S. Gupta, E.A. Fox, K. Bisset, and M.V. Marathe. 2014. Data mapping framework in a digital library with computational epidemiology datasets. In *Proceedings of the IEEE/ACM Joint Conference on Digital Libraries (JCDL).* London, 449–450.

International Telecommunication Union, Recommendation ITU-T Y.2060. 2012a. Overview of the internet of things.

International Telecommunication Union. 2012b. The state of broadband: Achieving digital inclusion for all. Broadband Commission for Digital Development Technical Report, September 2012.

Leidig, J., E. Fox, K. Hall, M. Marathe, and H. Mortveit. 2011a. Improving simulation management systems through ontology generation and utilization. In *Proceedings 11th annual international ACM/IEEE joint conference on Digital libraries,* 435–436.

Leidig, J., E. Fox, K. Hall, M. Marathe, and H. Mortveit. 2011b. SimDL: A model ontology driven digital library for simulation systems. In *Proceedings of the ACM/IEEE Joint Conference on Digital Libraries,* 81–84.

Leidig, J., E. Fox, and M. Marathe. 2011c. Simulation tools for producing metadata description sets covering simulation-based content collections. In *International Conference on Modeling, Simulation, and Identification.* p. 755(045).

Marathe, M., H. Mortveit, N. Parikh, and S. Swarup. 2014. Prescriptive analytics using synthetic information. *Emerging Methods in Predictive Analytics: Risk Management and Decision Making*. IGI Global.

Moulds, R. The internet of things and the role of trust in a connected world. The Guardian, January 23, 2014. Accessed June, 2014. http://www.theguardian.com/media-network/media-network-blog/2014/jan/23/internet-things-trust-connected-world.

Nebot, A., F.E. Cellier, and M. Vallverdu. 1998. Mixed quantitative/qualitative modeling and simulation of the cardiovascular system. *Computer Methods and Programs in Biomedicine* 55 (1998): 127–155.

Nicolescu, G., H. Boucheneb, L. Gheorghe, and F. Bouchhima. Methdology for efficient design of continuous/discrete-events co-simulation tools. In *Proceedings of 2007 Western Simulation Multiconference,* 172–179. ISBN 1-56555-311-X.

Parikh N., M. V. Marathe, and S. Swarup. 2016. Simulation summarization: (Extended abstract). In *Proceedings of the 2016 International Conference on Autonomous Agents & Multiagent Systems (AAMAS '16)*. Richland, SC, 1451–1452.

Tsoi, H. and W. Luk. 2011. FPGA-based smith-waterman algorithm: analysis and novel design. In *Proceedings of the 7th international conference on Reconfigurable computing: architectures, tools and applications,* 181–192. Springer-Verlag Berlin, Heidelberg. ISBN: 978-3-642-19474-0.

Woods, L. 2014. FPGA-enhanced data processing systems. Ph.D. Dissertation, ETH ZURICH.

Chapter 5
Uncertainty in Modeling and Simulation

**Wei Chen, George Kesidis, Tina Morrison, J. Tinsley Oden,
Jitesh H. Panchal, Christiaan Paredis, Michael Pennock,
Sez Atamturktur, Gabriel Terejanu and Michael Yukish**

5.1 Mathematical Foundations of Uncertainty in M&S

Modeling and simulation of complex real-world processes has become a crucial ingredient in virtually every field of science, engineering, medicine, and business. The basic objectives of model development are commonly twofold. The first objective is to explain from a scientific perspective relationships between independent/controllable model input variables and dependent responses or other

W. Chen
Northwestern University, Evanston, USA
e-mail: weichen@northwestern.edu

G. Kesidis
Pennsylvania State University, State College, USA
e-mail: kesidis@gmail.com

T. Morrison
Food and Drug Administration, Silver Spring, USA
e-mail: Tina.Morrison@fda.hhs.gov

J.T. Oden
University of Texas at Austin, Austin, USA
e-mail: oden@ices.utexas.edu

J.H. Panchal (✉)
School of Mechanical Engineering, Purdue University,
585 Purdue Mall, West Lafayette, IN 47907-2088, USA
e-mail: panchal@purdue.edu

C. Paredis
Georgia Institute of Technology, Atlanta, USA
e-mail: chris.paredis@me.gatech.edu

M. Pennock
Stevens Institute of Technology, Hoboken, USA
e-mail: mpennock@stevens.edu

© Springer International Publishing AG (outside the USA) 2017
R. Fujimoto et al. (eds.), *Research Challenges in Modeling
and Simulation for Engineering Complex Systems*, Simulation Foundations,
Methods and Applications, DOI 10.1007/978-3-319-58544-4_5

quantities of interests (QOIs). The second objective is to use the models and simulation for prediction and decision making. Whether used for explanation or prediction, a model can never fully explain (partially observed) past events or predict future events. It is, therefore, of central importance to understand the epistemological limitations of models and the uncertainty inherent in their predictions.

It is recognized that probability theory is the only theory of uncertainty consistent with the established normative tenets of epistemology. Probability is defined as a finite measure on a general space. Normative principles, which underlie the Kolmogorov axiomatization of probability measures (Aleksandrov et al. 1999), provide the philosophical and analytical justification that will agree with the tenets of mathematical analysis. Thus, probability theory is unique and serves the specific purpose of quantifying uncertainty via the measure theory that underlies modern philosophy and mathematics. Non-probabilistic approaches such as those based on intervals, fuzzy sets, or imprecise probabilities lack a consistent theoretical and philosophical foundation.

Even when researchers agree on Bayesian probability theory as a mathematical foundation for uncertainty in M&S, there are differences in the philosophical interpretation of this theory. In the engineering and M&S communities, much emphasis has been placed on Validation and Verification (V&V) and Uncertainty Quantification (UQ) (ASME 2006; NAS 2012). V&V consists of activities that aim to ascertain whether a model is "correct" and "credible." Similarly, UQ aims to quantify the uncertainties inherent in using a model and examine how these uncertainties are reflected in the model's predictions. The treatment of uncertainty in the engineering and M&S community often starts from a perspective of objective probability, which is inconsistent with the Bayesian perspective. Such treatment stems from the premise that uncertainty can be quantified objectively, and that such quantification can be based on data from repeated (past) experiments. From a Bayesian probability theory perspective, the notion of a probability as the limit of a relative frequency in a large number of repeated experiments cannot be justified. This is particularly the case when "extrapolation" is needed in prediction or when data about experiments cannot be collected, for instance, as in an engineering design context in which human and economic factors play an important role, and the artifact being designed does not yet exist. Instead, it is important to recognize that probabilities express subjective beliefs. Even when these beliefs are sometimes

S. Atamturktur
Clemson University, Greenville, USA
e-mail: sez@clemson.edu

G. Terejanu
University of South Carolina, Columbia, USA
e-mail: TEREJANU@cse.sc.edu

M. Yukish
Pennsylvania State Applied Research Laboratory, State College, USA
e-mail: may106@arl.psu.edu

strongly informed by large amounts of related and relevant data, and when this data has been incorporated through Bayesian updating, they remain an expression of subjective beliefs.

Even though UQ practice in engineering is gradually being influenced and enriched by rigorous methods drawn from modern statistics, probability, and philosophy, and the problem is increasingly being approached from a Bayesian perspective (Kennedy and O'Hagan 2001; Oden et al. 2017), some existing V&V and UQ literature appears to stylize ad hoc methods and lacks a consistent theoretical and philosophical foundation. One such example is the identification of uncertainty in the UQ and V&V literature in promoting a taxonomy that includes various definitions of both *aleatory* and *epistemic* uncertainty. Drawing distinction between aleatory and epistemic uncertainty for the purposes of predictive modeling is important only if the modeler seeks to employ classical Fisherian (frequentist) methods where the quantification of aleatory uncertainty is the focus. Over the last five decades, Fisherian methods, which are philosophically aligned with objective probability as interpreted by von Mises, have been largely replaced by Bayesian approaches. This overcomes the philosophical shortcomings of classical methods (e.g., Bayesian confidence replaces Fisherian hypothesis testing). From a Bayesian perspective, there is no need to delineate between aleatory and epistemic uncertainty as both are readily characterized within modern probability theory. Similarly, the UQ and V&V literature has introduced various ad hoc metrics of model validity. Since predictive models are developed as an aid to decision making, it would seem logical to define the notion of validity in a decision-making context. One would then base the treatment of uncertainty on value rather than validity as is further elaborated in Sect. 5.2.

Although probability theory provides the foundation for dealing with uncertainty, there are practical challenges in the application of probability theory in M&S. While physical processes (e.g., thermal, electrical, chemical, mechanical, multi-physics) are understood through models constructed from the laws of physics, our "certainty" about predicting the fidelity of physics-based models has no such foundation—no "physics of uncertainty" exists. Instead, uncertainty is a subjective assertion. The only physics-based assertions available for the characterization of uncertainty are "independence" (often stated through "conditional independence") and stationarity. Independence (along with stationarity) is a key ingredient of those few stochastic processes that yield to analysis (e.g., Brownian motion, Levy processes, regenerative processes, Markov processes, extreme value processes, branching processes) for which we have (perhaps functionally retrievable) probability laws. However, independence is rarely justified in engineered systems. This becomes a more daunting issue in M&S of complex systems, where sophisticated physics-based models when viewed as trajectories in a stochastic domain have insufficient independence structure to yield a functional characterization of probability law.

In summary, there is a need to unify behind a Bayesian perspective on uncertainty in M&S. Currently, activities such as model calibration, validation, uncertainty propagation, experimental design, model refinement, and decision making under uncertainty are studied in isolation and seen as an afterthought to model development. While agreeing on Bayesian probability theory as a foundation and on the mathematical models that should be applied in practice, there were philosophical differences regarding the meaning of probability. This has led to a significant fragmentation in the community where new ad hoc algorithms and models are continually proposed, but result in very little reuse by other practitioners. It is, therefore, important for the M&S community to be well-informed consumers of the available probability theory (as developed and accepted by philosophy, mathematics, and science) as opposed to developing competing methods that are at best redundant with existing methods or at worst invalid.

5.2 Uncertainty in the Context of Decision Making

Most often, M&S is ultimately used to guide decisions in engineering, medicine, policy, and other areas. Systems engineers make risk-informed design decisions. Medical professionals consider uncertainty when designing treatment strategies. Uncertain climate models influence policy decisions. It is clear that consistent consideration of uncertainty results in better decisions.

If model development were carried out independently of its eventual use in decision making, then one would conclude that reducing uncertainty is always better. In such a scenario, the modeler develops the best possible model under the budget constraints, and provides it to the decision maker, whose task is to bring in value judgments and make a decision using the available model. However, this is an inefficient approach for addressing uncertainty. It is explicitly recognized that M&S must be considered in the eventual context of use (COU), which defines the role and scope of the model in the decision-making process, and the resources available. Modelers must decide how much effort and resources to expend based on the potential impact on the decision. Based on the impact, they may then decide to refine the model with the goal of reducing uncertainty. In making these decisions, the role of the modeler is not simply to quantify model uncertainty, but to *manage uncertainty*. Uncertainty management (UM) is a broader activity in M&S that involves (i) establishing the decision COU and the modeling goals, (ii) identifying the effects of uncertainty on the decision, (iii) determining options for reducing uncertainty and the associated cost, (iv) evaluating the effects of these options on the goals, and (v) use of rigorous techniques to make consistent modeling decisions.

M&S processes are purpose-driven activities. The value of a modeling activity depends on the specific COU for which a model is developed. If the goal is to support selection among different alternatives, then information should be gathered only to the point that the best alternative can be determined. Within engineering design, for example, the primary COU of models is to help designers select among multiple design alternatives that maximize the designer's value. Therefore, the choice of model fidelity is dependent on designer's values, which are quantified by his/her preference function.

Various approaches have been well established within decision theory for modeling preferences. One of the approaches for quantifying preferences and value trade-offs under risk and uncertainty utilizes utility theory (Keeney and Raiffa 1993), which is based on axioms initially presented by von Neumann and Morgenstern (von Neumann and Morgenstern 1944). Utility theory forms the basis for rigorous approaches for estimating preference structures based on principles such as certainty equivalence. It is a foundation of microeconomics and is increasingly being adopted by many application domains such as engineering design and systems engineering.

Although utility theory can be used to formalize the preferences of a decision maker for whom a model is being developed, this can be challenging due to a number of reasons. First, the model developer may not have direct access to the decision maker or his/her preference structure during the model development process. This is particularly true when a model, initially developed to support one decision, is used for a different decision. Second, the preferences of the decision maker may evolve over time, or as more information becomes available. Third, modeling efforts may be driven by multiple uses and hence multiple target decisions. Fourth, modeling efforts may sometimes be driven by external factors, unrelated to the target decision. These challenges prevent direct application of the existing approaches and need further investigation by the research community.

After quantifying the preference structure, the next step is to evaluate the impact of modeling choices on the target decision. Models can improve the value of decisions, but also incur cost. They require time as well as computational, monetary, and human resources. The trade-off between the value of a model and associated costs leads to a broad question: *How much effort should be put into modeling activities to support decision making?* This question can be addressed by modeling uncertainty management itself as a decision-making process. Different models, at different fidelities, result in different accuracies and costs. The choice of model will have an impact on the expected outcome of the decision. More accurate model predictions tend to lead to more valuable decision outcomes but also cost more. If the increase in the expected value or utility of the decision outcome is larger than the expected cost of modeling, the modeling activity is worth pursuing. One can think of this process as efficient information gathering: Choose the information source (e.g., a model and corresponding simulation) that maximizes the net value of information (Lawrence 1999).

While this criterion for making modeling decisions is easy to state, implementing it during the model development process may be challenging. The criterion requires a comparison of cost and improvement in the decision. It may be difficult to quantify both of these quantities in the same units. The cost is generally related to (i) cost of collecting data, (ii) computational effort required to characterize the uncertainty in the predictions, and (iii) cost of employing subject matter experts to provide uncertainty assessments. Methods are needed for combining these cost attributes within a single measure. In addition, other modeling choices that influence the expected value of information need to be made: deciding which model to sample from, deciding how much experimental data to gather, deciding whether to refine a model or not, choosing the level of model fidelity, selection of general modeling approach (e.g., continuous simulation, agent-based simulation), deciding a model validation strategy, deciding whether to reuse existing models or to develop new customized models, choosing the level of abstraction for a model, deciding which multiscale models to integrate, deciding to compose models at different scales, etc. Each decision can be modeled from the perspective of value of information maximization. This general strategy has been utilized in techniques such as Bayesian global optimization based on sequential information acquisition (Jones et al. 1998), for model selection decisions (Moore et al. 2014), and for making different model calibration and verification decisions. Bayesian approaches are gaining particular importance due to the ability to integrate different sources of uncertainties (parameters and data) and to incorporate prior knowledge (Farrell et al. 2015).

The decisions listed above are usually made sequentially rather than in a single step. For example, the level of abstraction of the model is chosen before specifying the details of the parameter values. The decision-making process can be modeled as a decision network, where different decisions may be made by different individuals, perhaps within different teams, or even different organizations. In addition to the individual decisions, the structure of the decision network also affects the outcomes. The key question from the uncertainty management standpoint is: *How can resources for M&S activities be allocated efficiently to maximize the value gained from the network of decisions?* This is itself a computationally challenging, dynamic decision-making problem.

The organizational context in which M&S activities are framed presents additional challenges. For instance, M&S may be part of the systems engineering process, which in turn is a part of the overall business process. Therefore, modeling activities may compete for resources with many other activities within the organization, or time constraints may be imposed based on external factors such as market competition. There is, therefore, a need to establish techniques for partitioning the budget for different activities and targets within the context of organizational goals.

In summary, there are three key decision-related research challenges in M&S. The first challenge is to consistently deduce the preference functions for individual uncertainty management activities from overall goals within organizations where multiple entities are involved in decision making and their preferences may be conflicting. The second is related to mapping uncertainty in physical quantities to the utility functions. The third is due to the complexity of sequential decision-making processes with information acquisition. Addressing these challenges would help in partitioning and allocating organizational resources for modeling and decision making under uncertainty.

5.3 Aggregation Issues in Complex Systems Modeling

Aggregation of information is an integral part of M&S of complex systems. The common approach for managing complexity is to follow a divide-and-conquer strategy, which involves partitioning the modeling activity based on various criteria, such as type of physical phenomena, level of detail, expertise of individuals, and organizational structure. The models of the partitioned system are then integrated into system-level models to provide a holistic representation of the system behavior. Based on the criteria used for partitioning the modeling task, the techniques are referred to as multi-physics, multi-disciplinary, multi-fidelity, and multi-scale techniques. These techniques are gaining increasing attention in many application domains ranging from computational materials science to critical infrastructure design (Felippa et al. 2001). In computational materials science, for example, models are developed at multiple levels including continuum, meso- and micro-scales and atomistic levels.

Aggregation of information is associated with a number of challenges in M&S. First, composition of models requires an understanding of physical phenomena at different levels, and how they can be seamlessly integrated across different levels. This is referred to as scale bridging within the multiscale modeling literature, and strategies ranging from hierarchical to concurrent modeling are being developed (Horstemeyer 2010). Second, there is a need for rigorous approaches for modeling uncertainty across different scales. Since different individuals, teams, and organizations develop models, the sources of uncertainty and the domain of applicability of individual models may be different. Modeling assumptions may not be consistent across different models. These inconsistencies across models can result in erroneous predictions about the behaviors at the aggregate level.

A greater challenge in such divide-and-conquer strategies is that even if consistency across different models is achieved, the fundamental nature of aggregation can result in erroneous results due to the path dependency problem. Saari (2010) shows that multi-level methodologies can be treated as generalizations of aggregation processes. Although each lower-level model provides strong evidence for seemingly logical outputs, the conclusions at the aggregate level can be incorrect. The aggregate level output could merely reflect the way in which lower-level models are assembled rather than the actual system behavior (Saari 2010; Stevens and Atamturktur 2016). The primary cause of the inaccuracy is that separation caused by divide-and-conquer strategy loses information.

The potential inaccuracies resulting from aggregation have been studied in detail in relation to aggregation of preferences for group-decision making. It has been shown that aggregation procedures can result in biases in decisions (Saari and Sieberg 2004; Hazelrigg 1996). The implication for simulation-based design is that commonly used decision-making methods based on normalization, weighting, and ranking are likely to lead to irrational choices (Wassenaar and Chen 2003). This is particularly important in M&S because, as discussed in Sect. 5.2, the model development process involves many decisions made by different decision makers.

The aggregate system-level model represents an aggregation of beliefs and preferences of individual lower-level model developers. Therefore, the inaccuracies resulting from preference aggregation is a fundamental challenge in M&S of complex systems. In summary, there is a need to recognize and address the challenges associated with aggregation of physics-related and preference-related information in modeling complex systems.

5.4 Human Aspects in M&S

Human aspects in M&S are important for two reasons. First, humans are integral parts of socio-technical systems, such as electric power grids, smart transportation systems, and healthcare systems. Therefore, accurately modeling human behavior is essential for simulating the overall system behavior. Second, the developers and users of models are human decision makers. Therefore, the effectiveness of the model development and usage process is highly dependent on the behavior of the decision makers.

With the rapid rise of smart networked systems and societies (Simmon et al. 2013), which consist of people, internet-connected computing devices, and physical machines, modeling humans within the overall system has become an essential part of M&S activities. Within such cyber-physical-social systems, humans receive information over the network, interact with different devices, and make decisions that affect the state of the system. The key challenge in modeling such systems is to determine how to incorporate human behavior into formal models of systems. Modeling human behavior is challenging because of complex and uncertain physiological, psychological, and behavioral aspects. Humans are generally modeled with *attributes* such as age, sex, demographic information, risk tolerance, and *behaviors* such as product and energy usage. Such an approach is common in agent-based models. Another class of models is human-in-the-loop models, where humans are part of the simulation.

As discussed earlier, M&S is a decision-making process, and the decision makers are humans. Humans are known to deviate from ideal, rational behavior. For example, decision makers exhibit systematic biases in judgment of uncertainty (Kahneman et al. 1982), inconsistencies in preferences, and in the process of utilizing the process of expected utility theory. These deviations from normative models can be attributed to a number of factors such as cognitive limitations, performance errors, and incorrect application of the normative model (Stanovich 1999).

The gap between normative and descriptive models of human decision making has been well documented within the fields of behavioral decision research and psychology. Behavioral experiments have provided insights into how humans deviate from normative models, which have been used to develop psychological theories to explain these deviations. These deviations have been modeled in descriptive theories such as prospect theory, dual process theory, and many others.

Alternate theories about decision making based on simple heuristics have also been proposed (Gigerenzer et al. 1999). These heuristics extend from simple one-step decisions to multi-step decisions with information acquisition at each step. Behavioral studies have also been extended to interactive decisions modeled using game theory (Camerer 2003). Recently, psychologists have started exploring neuro-science as a way to understand human behavior in general, and decision making in particular (Camerer et al. 2005).

While there has been significant progress on understanding humans as decision makers, the utilization of this knowledge in M&S activities has been limited. There are a number of open questions such as (i) how do these biases affect the outcome of modeling decisions? (ii) how can these biases be reduced? (iii) how can the effects of these deviations from rationality be reduced within the M&S process? (iv) what is the best way of presenting and communicating uncertainty information to the decision makers? (v) what is the effect of domain-specific expertise and knowledge on deviations from rational behavior? and (vi) are there differences in biases between novice and expert modelers?

From an organizational standpoint, there are multiple individuals involved in the modeling process. Different individuals may have different beliefs, may be driven by different values, and may be influenced differently by different types of biases. These values, beliefs, and biases get embedded in their individual models. Further research is necessary to establish how these interact within an organization, to ensure consistency across values and beliefs, and to overcome biases of individuals. This is clearly not a comprehensive list, but it highlights the importance of considering human aspects in M&S and provides some pointers for further investigation.

In summary, human considerations are important for creating better models of systems involving humans, and for simulation of social-technical systems. Additionally, human considerations are important for better understanding of biases that exist during the modeling decisions made by humans. Addressing human aspects within M&S would help in designing better control strategies for smart networked systems and societies, better M&S processes, efficiently allocating organizational resources, and making better model-driven decisions. Research toward answering these questions would require collaboration between domain-specific modeling researchers and researchers in social, behavioral, and psychological sciences.

5.5 Communication and Education of Uncertainty in M&S

An additional challenge regarding uncertainty in M&S is: *how to effectively communicate model predictions among various stakeholders*? Especially in model reuse or when passing the models from model developers to decision makers, it is important to state clearly the key underlying assumptions along with their potential impact on the predicted QOIs. Sensitivity of key outcomes to the alternative modeling assumptions should also be assessed and presented effectively. Visualization tools need to be developed for illustrating the uncertainty sources, how they propagate, and their impacts over the entire domain of interest.

Related to the topic of communication of uncertainty is education for both students and faculty. At present, undergraduate students are typically taught the existing models related to each course subject without being introduced to the significance of the modeling process or a critical assessment of associated assumptions and uncertainties. For example, in engineering design courses, students are most often introduced to ad hoc approaches to deal with uncertainty, such as the use of "safety factors." Courses on probability and statistics are often elective but not required, and students often take advanced science and engineering courses before they have gained exposure to probability and statistics. Moreover, probability and statistics courses for engineering undergraduate students deal largely with data analysis and do not introduce many concepts that are important to prediction and decision making under uncertainty.

A modern curriculum on probability in engineering and science is, therefore, needed to equip students with the foundation to reason about uncertainty and risks. A modern curriculum should foster an appreciation of the role that M&S could play in addressing complex problems in the interconnected world. The curriculum should also address effective communication of uncertainty and risk to modelers, decision makers, and other stakeholders.

5.6   Other Issues: Integration of Large-Scale Data

The advent of ubiquitous and easy-to-use cloud computing has more readily enabled simulations to leverage huge real-world datasets, i.e., "big data." Naturally, real-world datasets may be used to inform the models of the simulated system and its inputs, or could also be used to validate these models by comparing output behavior with real-world observations. A challenge of big real-world datasets is that they may be incomplete or noisy and include samples taken in different contexts so that they need to be "detrended" based on covariate information (contextual metadata). Moreover, many or most of the sample features available may be superfluous to the simulation objectives. Noisy and superfluous features may result in inaccurate (e.g., overfitted) and needlessly complex models, again considering the specific simulation objectives. Techniques have been developed by data scientists to reduce or combine features of a sample dataset, e.g., using the classical methods of multi-dimensional scaling or principal component analysis. Future work in this area includes model-specific techniques of feature selection. Note that large datasets may not only have large numbers of samples but samples with enormous numbers of features (high feature dimension), so that future work in this area also includes scalable (low complexity) and adaptive techniques, the latter for dynamic, time-varying settings.

Large-scale datasets are often also used for deriving complex empirical relationships, using machine learning algorithms. Overfitting is a common concern in such scenarios. Cross-validation or hold-out testing provides a direct demonstration of a model's ability to predict under new conditions not encountered in the training set. Such methods use a majority of the data to calibrate or correct a model while holding some data to predict experimental or observational outcomes that were not used in the model calibration process. Characterizing uncertainty in extrapolative settings and rare events are challenging topics that require new research approaches that incorporate rigorous mathematical, statistical, scientific, and engineering principles.

References

Aleksandrov, A. D., A. N. Kolmogorov, and M. A. Lavrent'ev. 1999. *Mathematics: Its Content, Methods and Meaning*. Courier Corporation.

ASME Committee PTC-60 V&V 10. 2006. Guide for verification and validation in computational solid mechanics. https://www.asme.org/products/codes-standards/v-v-10-2006-guide-verification-validation.

Camerer, C.F. 2003. *Behavioral Game Theory: Experiments in Strategic Interaction*. Princeton University Press.

Camerer, C.F., G. Loewenstein, and D. Prelec. 2005. Neuroeconomics: how neuroscience can informeconomics. *Journal of Economic Literature* 43 (1): 9–64.

Farrell, K., J.T. Oden, and D. Faghihi. 2015. A Bayesian framework for adaptive selection, calibration, and validation of coarse-grained models of atomistic systems. *Journal of Computational Physics* 295: 189–208.

Felippa, C.A., K.C. Park, and C. Farhat. 2001. Partitioned analysis of coupled mechanical systems. *Computer Methods in Applied Mechanics and Engineering* 190 (24): 3247–3270.

Gigerenzer, G., P. Todd, and A. Group. 1999. *Simple Heuristics that Make us Smart*.

Hazelrigg, G.A. 1996. The implication of arrow's impossibility theorem on approaches to optimal engineering design. *ASME Journal of Mechanical Design* 118: 161–164.

Horstemeyer, M.F. 2010. Multiscale modeling: a review. In *Practical Aspects of Computational Chemistry: Methods, Concepts and Applications*. Chap. 4, Springer Netherlands, 87–135.

Jones, D.R., M. Schonlau, and W.J. Welch. 1998. Efficient global optimization of expensive black-box functions. *Journal of Global Optimization* 13: 455–492.

Kahneman, D., P. Slovic, and A. Tversky. 1982. *Judgment Under Uncertainty: Heuristics and Biases*. Cambridge University Press.

Keeney, R.L., and H. Raiffa. 1993. *Decisions with Multiple Objectives: Preferences and Value Tradeoffs*. Cambridge: UK, Cambridge University Press.

Kennedy, M.C., and A. O'Hagan. 2001. Bayesian calibration of computer models. *Journal of the Royal Statistical Society Series B-Statistical Methodology*. 63: 425–450.

Lawrence, D.B. 1999. *The Economic Value of Information*. New York, NY: Springer.

Moore, R.A., D.A. Romero, and C.J.J. Paredis. 2014. Value-based global optimization. *Journal of Mechanical Design* 136 (4): 041003.

National Academy of Science (NAS). 2012. Assessing the reliability of complex models: mathematical and statistical foundations of verification, validation, and uncertainty quantification. *NAS Report*. doi:10.17226/13395.

Oden, J. T., I. Babuska, D. Faghihi. 2017. Predictive computational science: computer predictions in the presence of uncertainty. In *Encyclopedia of Computational Mechanics*, Wiley and Sons (to appear).

Saari, D.G., and K. Sieberg. 2004. Are part wise comparisons reliable? *Research in Engineering Design* 15: 62–71.

Saari, D.G. 2010. Aggregation and multilevel design for systems: finding guidelines. *Journal of Mechanical Design* 132 (8): 081006.

Simmon, E.D., K-S. Kim, E. Subrahmanian, R. Lee, F.J. deVaulx, Y. Murakami, K. Zettsu, and R. D Sriram. 2013. A Vision of Cyber-Physical Cloud Computing for Smart Networked Systems. NISTIR 7951. http://www2.nict.go.jp/univ-com/isp/doc/NIST.IR.7951.pdf.

Stanovich, K.E. 1999. *Who Is Rational? Studies of Individual Differences in Reasoning*. Mahwah, NJ: Lawrence Erlbaum Associates Inc.

Stevens, G. and S. Atamturktur. 2016. Mitigating error and uncertainty in partitioned analysis: a review of verification, calibration and validation methods for coupled simulations. *Archives of Computational Methods in Engineering*, 1–15.

von Neumann J. and O. Morgenstern. 1944. *Theory of Games and Economic Behavior*. Princeton University Press.

Wassenaar, H.J., and W. Chen. 2003. An approach to decision-based design with discrete choice analysis for demand modeling. *ASME Journal of Mechanical Design* 125 (3): 490–497.

Chapter 6
Model Reuse, Composition, and Adaptation

Osman Balci, George L. Ball, Katherine L. Morse, Ernest Page, Mikel D. Petty, Andreas Tolk and Sandra N. Veautour

The motivations for the reuse of models are well founded. Models are *knowledge artifacts*, and as such, their reuse provides the opportunity for scientists, engineers, and educators to "stand on the shoulders of giants." Models are also typically manifest as *software* that has been developed with significant effort and subjected to rigorous testing and verification and validation. The attractiveness of the potential cost and labor savings associated with the reuse of this software is quite understandable.

O. Balci
Virginia Tech, Blacksburg, VA, USA
e-mail: balci@vt.edu

G.L. Ball
Raytheon Company, Waltham, MA, USA
e-mail: george_ball@raytheon.com

K.L. Morse
Applied Physics Laboratory, The Johns Hopkins University, Laurel, MD, USA
e-mail: Katherine.Morse@jhuapl.edu

E. Page (✉)
The MITRE Corporation, 7515 Colshire Drive, Mclean, VA 22102, USA
e-mail: epage@mitre.org

M.D. Petty
University of Alabama in Huntsville, Huntsville, AL, USA
e-mail: pettym@uah.edu

S.N. Veautour
U.S. Army Aviation and Missile Research Development and Engineering Center,
Huntsville, AL, USA
e-mail: Sandra.veautour@mda.mil

A. Tolk (✉)
The MITRE Corporation, 903 Enterprise Parkway, Hampton, VA 23666, USA
e-mail: atolk@mitre.org

© Springer International Publishing AG (outside the USA) 2017
R. Fujimoto et al. (eds.), *Research Challenges in Modeling and Simulation for Engineering Complex Systems*, Simulation Foundations, Methods and Applications, DOI 10.1007/978-3-319-58544-4_6

The reuse of models is confounded, however, by the fact that they are peculiarly *fragile* in a certain sense—they are purposeful abstractions and simplifications of a perception of a reality. This perception has been shaped under a possibly unknown set of physical, legal, cognitive, and other kinds of constraints and assumptions. The end result is that model reuse tends to be much more challenging than, say, reusing the implementation of a sorting routine.

While some communities of practice (e.g., micro-electronics design, defense training) can arguably be viewed as success stories in the development and adoption of both the technologies and business practices for model reuse, general solutions to this important problem remain elusive.

There has been much significant work in the area of model reuse and in software reuse more broadly. While it is beyond the scope of this effort to provide a comprehensive survey, several notable works are cited herein. In this report, we focus on three distinct areas for recommended further study:

- *Advancements in the theory of reuse.* Without a firm theoretical foundation, we cannot fully know the fundamental limits of what we can hope to accomplish with reuse. Properly formulated, good theory may also be exploited to produce robust and reliable reuse practices.
- *Advancement in the practice of reuse.* In this context, we consider the following: (1) modeling and simulation (M&S) broadly, (2) data, and (3) discovery and knowledge management.
- *Advancements in the social, behavioral, and cultural aspects of reuse.* Here, we consider how incentives may stimulate or impede reuse.

6.1 Advancements in the Theory of Reuse

Reuse has been defined as "Using a previously developed asset again, either for the purpose for which it was originally developed or for a new purpose or in a new context" (Petty et al. 2010) and *reusability* as "the degree to which an artifact, method, or strategy is capable of being used again or repeatedly" (Balci et al. 2011). In the former definition, an *asset* is "a reusable collection of associated artifacts." Assets may be software components, data sets, documentation, design diagrams, or other development artifacts, but for brevity and simplicity, we use the term here primarily to refer to software components.

In the context of modeling and simulation, an asset is a software component that implements all or part of a model (e.g., a software component that implements a physics-based model of a jet aircraft engine) or all or part of the software needed to *support* a model (e.g., a component that implements an XML-based scenario initialization operation). When a distinction is needed, the former category will be referred to as *model components* and the latter as *support components*.

Metadata is supplemental information about a component that may be used for a number of purposes. In modeling and simulation, a model component's metadata may describe the model's function, intended use, assumptions, and uncertainties, in a way that enables appropriate reuse and reduces inappropriate reuse of the component (Taylor et al. 2015).

We observe that any theory of reuse for M&S does need to be formed from "whole cloth." Several theories from computing and mathematics support the development of a theory for M&S reuse, including: computability theory, computational complexity theory, predicate logic, algorithmic information theory, model theory, and category theory.

6.1.1 Prior Theoretical Work Relating to M&S Reuse

Past theoretical work relating to modeling and simulation reuse is briefly summarized, and several key results are described next.

6.1.1.1 Composability

Composability is the capability to select and assemble simulation components in various combinations into simulation systems to satisfy specific user requirements (Petty and Weisel 2003). Although composability and reusability are not the same idea (Balci et al. 2011; Mahmood 2013), composability can be an important enabler for reuse. For close to two decades, composability has been a focus for simulation developers, particularly in the defense-related modeling and simulation community. Composability was identified as a key objective as early as 1999 (Harkrider and Lunceford 1999) and was recently described as "still our biggest simulation challenge" (Taylor et al. 2015). Composability applies to both model components and support components, although much composability research has focused on mechanisms for composing models and the validity of the resulting composite models. Some notable results include the following:

- The development and community adoption of a common terminology and formal definitions for composability, as well as the related concepts integratability and interoperability (Petty and Weisel 2003; Petty et al. 2003a; Page et al. 2003; Tolk and Muguira 2003).
- Characterizations of the computational complexity of selecting models to be composed (Page and Opper 1999; Petty et al. 2003b).
- The development of a theoretical basis for characterizing the validity of any composition of separately validated model components (Weisel et al. 2003).
- Demonstration that a "simple form" of composition is sufficient to assemble any composite model (Petty 2004).

In some ways, this foundational work has established that composition provides no "silver bullet" for the fundamental complexities of modeling and simulation. For example, a common assumption made by simulation developers is that if two models have been separately determined to be valid, then those models (or the components implementing them) may be composed and the resulting composition will also necessarily be valid. Weisel et al. (2003) show that in all but the most trivial cases, the composition of two (or more) separately valid models cannot be assumed to be valid. Similarly, one can easily construct scenarios where individual models are invalid for a specific purpose; however, their composition creates a valid model (e.g., the component models are incomplete, or their "weaknesses" are offset by the other component model). The implication of this result is simple—composition provides no relief to the fundamental costs or complexities of model validation. Irrespective of the validity of the individual components, the overall composite model must always be validated separately.

Similarly, software developers have produced and will continue to produce sophisticated software frameworks for combining, or composing, models, with the intention of making such compositions easier to assemble and execute (Petty et al. 2014). It might be reasonable to assume that an increasing capability in these software frameworks would have a fundamental impact on what can and cannot be achieved in terms of model composability. Petty (2004) illustrates that all composed models can be described in terms of a "simple" underlying composition formula, and therefore, the manner of composition does not fundamentally alter the nature of the composed model.

6.1.1.2 Component Selection

Component selection is the computational problem of selecting from a repository containing a set of available components a subset of those components to be composed so that the resulting composition will satisfy a given set of objectives for a simulation system (Clark et al. 2004). Note that there are actually two computation problems in component selection. The first problem is to *determine* which requirements a component satisfies, either in advance of component selection or on request when a set of requirements are presented. The second problem is to *select* a set of components to meet a given set of requirements. Both of these problems are well-known in software engineering. Pressman and Maxim (2015) summarize them as "How do we describe software components in unambiguous, classifiable terms?" and "[H]ow do you find the [components] that you need?" respectively.

To select a group of components that collectively meet some set of objectives, the objectives met by each component must be determined. Unfortunately, it is easy to see that such a determination may be problematic. Suppose a desired objective for a component is that it complete execution (rather than enter an infinite loop) for all inputs. This is the well-known "halting problem," which is known to be incomputable in the general form. Even objectives that, in principle, can be algorithmically decided may require a computation time that is superpolynomial and thus infeasible in practice (Page and Opper 1999). The implication of this result is that the determination of the objectives met by a component may have to be done by means other than purely algorithmic, no matter how sophisticated the approach to component metadata.

Even if the objectives met by each component in the repository are somehow known, component selection remains difficult. In Page and Opper (1999), the specific form of component selection most similar to the practical application was shown to be NP-complete by reduction from SATISFIABILITY. In Petty et al. (2003a), a general form of component selection was shown to be NP-complete by reduction from MINIMUM COVER. The implication of this result is the same as any NP-complete problem; the computational problem (in this case component selection) cannot be solved algorithmically in general, and heuristics that produce acceptable selections for most instances of the problem will have to be developed.

6.1.2 Research Topics in M&S Reuse Theory

Three research topics related to advancing the theory of modeling and simulation reuse and applying that theory in practical settings are suggested. They are listed from "most theoretical" to "most practical," and a set of relevant research questions for each topic is listed.

6.1.2.1 Composability Theory

Understanding the theoretical limits of composability, i.e., the composition of models and the validity of such compositions, is essential. Work has been started, but a fully coherent, comprehensive, and mature theory of composability has not yet been developed. Relevant research questions include the following:

1. *What are the theoretical characteristics or attributes of a model or component composition, beyond simply a composition of computable functions, and how do they affect reuse?*
2. *What theoretical formalisms are most effective at describing and analyzing composability?*
3. *Can models at different levels of abstraction or based on different modeling paradigms be composed without loss of validity?* (Fujimoto 2016)
4. *Can the operations and problems of modeling and simulation reuse be recast in the terms and concepts of algorithmic information theory, category theory, and model theory, and if so, what insights would that provide?*
5. *Although as noted earlier the overall validity of a model composition is not assured simply by the validity of the model components, can anything about the validity of the composition be inferred from the components and the way in which they are composed?* (Tolk et al. 2013)

6.1.2.2 Metadata and Reuse

Metadata is often described as enabling reuse, and a rigorous theoretical approach to metadata is conjectured to be more likely to succeed that ad hoc specifications. Predicate logic is arguably most often among the formalisms proposed for metadata, but it has not yet been demonstrated to be usable in practical settings. Relevant research questions include the following:

1. *What formalism(s) are suitable for expressing component metadata?*
2. *What characteristics of a model should be expressed in metadata?*
3. *Can developing a standard vocabulary, perhaps defined using some form of ontology, increase the effectiveness of metadata?*
4. *Can component metadata be algorithmically or heuristically generated from or verified against a component?*
5. *How can the assumptions made in a model be expressed in metadata and used in component selection and model composition?*

6.1.2.3 Reuse Automation

Algorithms and frameworks that automate reuse operations, including component selection and composition verification and validation, would likely expand the frequency and value of reuse. Relevant research questions include the following:

1. *How can/should model selection, composition, and code generation be supported by automated?* (Fujimoto 2016)
2. *What forms of theoretical composition correspond to practical reuse patterns?*
3. *Can reuse patterns themselves be reused, in the manner of design patterns?*
4. *Can the validity of a proposed composition of models be algorithmically confirmed?*
5. *Can heuristics be developed to circumvent theoretical obstacles and provide reasonable performance in most practical situations?*
6. *Can constraints imposed on model development (e.g., standards) improve the composability of the models once developed?* (Fujimoto 2016)

6.2 Advancements in the Practice of Reuse

In this section, we consider some of the challenges to the day-to-day practice of reuse in a modeling context. We separate the discussion into three distinct areas:

- Modeling and simulation—in which we deal with issues confronting the reuse of representations of models and their implementation in simulation languages and frameworks.

- Data—in which we deal with the issues confronting the reuse of those elements that are consumed and produced by models.
- Knowledge management and discovery—in which we address the issues involved in archiving and discovering artifacts (models, simulations, data) that may be reused.

6.2.1 Challenges in the Practice of Reuse of Models and Simulations

We identify research challenges associated with the reuse of model representations and their implementation as simulations in four areas: (1) multi-formalism, multi-scale modeling, (2) reuse across communities of interest and the implementation spectrum, (3) exploitation of M&S Web services, and (4) quality-centric approaches to component evaluation. Each of these is discussed below.

6.2.1.1 Multi-formalism, Multi-scale Modeling

As a topical matter, M&S is incredibly broad. It spans dozens of disciplines and countless potential objectives and intended uses (ACM SIGSIM 2016). To our knowledge, a definitive, exhaustive taxonomy for M&S has not been formulated. For the purposes of this report, we adopt the characterization given in Table 6.1. While necessarily incomplete, it is indicative of the breadth of M&S. Each area noted in the table possesses its own characteristics and methodologies, is applicable for solving certain classes of problem, and has its own community of users. Many M&S areas have their own societies, conferences, books, journals, and software tools.

The current era of "net-centricity" has resulted in a proliferation of "systems of systems" in which disparate systems with diverse characteristics are composed and integrated over networks, e.g., the Internet, virtual private networks, wireless networks, and local area networks.

We face serious technical challenges in achieving reusability, composability, and adaptability for developing simulation models representing such network-centric systems of systems. Different systems or system components may be required to be modeled by using different M&S types and/or at vastly different spatial and temporal scales. For example, one component may be modeled using discrete M&S, another using Computational Fluid Dynamics, another in Finite Element, and still another using System Dynamics. Achieving interoperability across these modeling approaches is an open problem.

New methodologies, approaches, and techniques must be created to enable the development of an M&S application or component by way of reusing, composing, and adapting different types of M&S applications or components.

Table 6.1 M&S areas (types) (Balci, Introduction to Modeling and Simulation 2016; Balci et al. Achieving Reusability and Composability with a Simulation Conceptual Model 2011)

A.		Based on model representation	Development approach
	1.	Discrete M&S	Logic
	2.	Continuous M&S	Differential equations
	3.	Monte Carlo M&S	Statistical random sampling
	4.	System Dynamics M&S	Rate equations
	5.	Gaming-based M&S	Logic
	6.	Agent-based M&S	Knowledge, "intelligence"
	7.	Artificial intelligence-based M&S	Knowledge, "intelligence"
	8.	Virtual reality-based M&S	Computer generated visualization
B.		Based on model execution	
	9.	Distributed/Parallel M&S	Distributed processing/computing
	10.	Cloud-based M&S	Cloud software development
C.		Based on model Composition	
	11.	Live exercises	Synthetic environments
	12.	Live experimentations	Synthetic environments
	13.	Live demonstrations	Synthetic environments
	14.	Live trials	Synthetic environments
D.		Based on what is in the loop	
	15.	Hardware-in-the-loop M&S	Hardware + simulation
	16.	Human-in-the-loop M&S	Human + simulation
	17.	Software-in-the-loop M&S	Software + simulation

6.2.1.2 Artifact Reuse Across Communities of Interest and the Implementation Spectrum

Many different types of M&S applications are commonly employed in a community of interest (COI) such as air traffic control, automobile manufacturing, ballistic missile defense, business process reengineering, emergency response management, homeland security, military training, network-centric operations and warfare, supply chain management, telecommunications, and transportation. Reusability, composability, and adaptability are critically needed to facilitate the design of any type of large-scale complex M&S application or component, in a particular COI, and significantly reduce the time and cost of development.

An M&S application or component is developed in a COI under a certain terminology (e.g., agent, job, missile). In another COI, the same M&S application or component may be developed from scratch without any kind of reuse because the terminology does not match although they are basically the same applications or components.

Challenges in reuse across the wide spectrum of implementations are also important. In M&S application development, we should aim to reuse, compose, and/or adapt an artifact, development process, design pattern, or framework such as:

Fig. 6.1 Levels of reusability versus achievability (Balci et al. 2011)

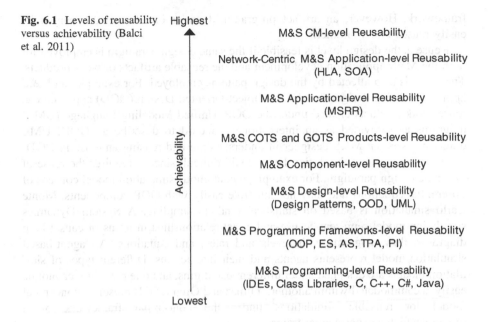

(1) a simulation program subroutine, function, or class, (2) a simulation programming conceptual framework, (3) a simulation model/software design pattern, (4) a simulation model component or submodel, (5) an entire simulation model, or (6) conceptual constructs for simulation in a particular problem domain.

Figure 6.1 depicts how well reusability can be achieved at different levels of M&S application development.

At the programming level, classes (under the object-oriented paradigm) and subroutines/functions (under the procedural paradigm) are extracted from a library using an Integrated Development Environment (IDE) such as Eclipse, NetBeans, or Microsoft Visual Studio. However, reuse at this level is extremely difficult due to the many options in programming languages (e.g., C, C#, C++, Java), differences in operating systems (e.g., Unix, Windows), and variations among hardware platforms (e.g., Intel, SPARC, GPU, FPGA) supporting language translators. An artifact programmed in Java and executing under a Unix operating system on a SPARC workstation cannot be easily reused in an M&S application being developed in C++ under the Windows Operating System on an Intel-based workstation.

M&S programming frameworks may be categorized according to the underlying programming paradigm, e.g., object-oriented paradigm (OOP), procedural paradigm (PP), functional paradigm (FP), and so forth. Balci (1998) describes four conceptual frameworks under the PP for simulation programming in a high-level programming language: event scheduling (ES), activity scanning (AS), three-phase approach (TPA), and process interaction (PI). A simulation programmer is guided by one of these frameworks by reusing the concepts supported in that conceptual framework. However, an artifact programmed under one framework cannot be easily reused under another.

Reuse at the design level is feasible if the same design paradigm is employed for both the M&S application development and the reusable artifacts or work products. The reuse is also affected by the design patterns employed. For example, an M&S application being designed under the Object-Oriented Design (OOD) approach can reuse work products created under the OOP. Unified Modeling Language (UML) diagrams are provided as an international standard to describe an OOD. UML diagrams assist an M&S designer in understanding and reusing an existing OOD.

However, reuse at the design level is still difficult since it requires the reuse of the same design paradigm. For example, a continuous simulation model consists of differential equations and may not integrate easily with OOP components. Monte Carlo simulation is based on statistical random sampling. A System Dynamics simulation model represents cause-and-effect relationships in terms of causal loop diagrams, flow diagrams with levels and rates, and equations. An agent-based simulation model represents agents and their interactions. Different types of simulation models are designed under different paradigms, and one paradigm cannot be easily accommodated within another. Yilmaz and Ören (2004) present a conceptual model for reusable simulations under the conceptual framework of a model-simulator-experimental frame.

M&S component level reuse is intended to enable the assembly (composition) of a simulation model by way of employing already developed model components in a similar fashion as an automobile is assembled from previously produced parts. A component may correspond to a submodel or a model module. Reuse at this higher level of granularity is beneficial because it reduces development time and cost over that of reuse at the class or function level. However, this approach to reuse still poses difficulties since each reusable component can be implemented in a different programming language intended to run under a particular operating system on a specific hardware platform.

M&S Commercial Off-The-Shelf (COTS) (e.g., Arena, AutoMod, and OpNet) and Government Off-The-Shelf (GOTS) products enable reuse of components within their IDEs. Such an IDE provides a library of reusable model components. A user can click, drag, and drop an already developed component from the library and reuse it in building a simulation model. However, such reuse is specific only to that particular COTS or GOTS IDE, and portability to another IDE would become a user's responsibility.

Reuse at the application level is feasible if the intended uses (objectives) of the reusable M&S application match the intended uses of the M&S application under development. For example, the US Department of Defense (DoD) provides the DoD M&S Catalog (Modeling and Simulation Coordination Office (MSCO) 2016) containing previously developed M&S applications. Some of these applications are independently certified for a set of intended uses. Some are not well documented and come in binary executable form only. Even if the source code is provided, understanding the code sufficiently well to modify the represented complex behavior is extremely challenging. Reusability of earlier developed M&S

applications is dependent on run-time environment compatibility and the match between intended uses.

A network-centric M&S application involves M&S components interoperating with each other over a network, typically for the purpose of accommodating geographically dispersed persons, laboratoriess, and other assets. The High-Level Architecture (HLA) is a DoD, IEEE, and NATO standard for developing network-centric M&S applications by way of interoperation of simulation models distributed over a network (IEEE, IEEE Standard 1516, 1516-1, 1516-2, and 1516-3). If a simulation model is built in compliance with the HLA standard, then that model can be reused by other models interconnected through the HLA protocol over a network.

Service-Oriented Architecture (SOA) is yet another architecture based on the industry standard Web services and the eXtensible Markup Language (XML). SOA can be employed for developing a network-centric M&S application by way of reuse of simulation models, submodels, components, and services over a network. For example, Sabah and Balci (2005) provide a Web service for random variate generation (RVG) from 27 probability distributions with general statistics, scatter plot, and histogram of the requested random variates. The RVG Web service can be called from any M&S application that runs on a server computer over a network using XML as the vehicle for interoperability. Reuse, composability, and interoperability are fully achieved regardless of the programming language, operating system, or hardware platform. However, this type of reuse is possible only for network-centric or Web-based M&S application development.

New methodologies, approaches, and techniques must be created to enable the development of an M&S application in a COI by way of reusing, composing, and adapting other M&S applications or components created in other COIs. New methodologies, approaches, and techniques are needed to enable the development of an M&S application through reuse, composition, and adaptation of M&S applications or components across the spectrum of implementation levels.

6.2.1.3 M&S Web Services

This research challenge deals with *how* to reuse, compose, or adapt. Initiated in the early 2000s, the US National Institute of Standards & Technology (NIST) Advanced Technology Program (ATP) cited many advantages of component-based development that could be realized conditioned on the following (NIST 2005):

(1) Establishment of a marketplace for component-based software development so that the technology users can realize significant economic benefits through

   (a) reduced software project costs,
   (b) enhanced software quality, and
   (c) expanded applicability of less expensive technology.

(2) Increased automation and productivity in software development enabling

  (a) improved software quality characteristics,
  (b) reduced time to develop, test, and certify software, and
  (c) increased amortization of costs through software component reuse.

(3) Increased productivity of software project teams by

  (a) permitting specialists in the application domain to create components incorporating their expertise and
  (b) providing a focus on discourse in development at a level far more higher level than a programming language.

(4) Expanded markets for software applications and component producers by promoting

  (a) the creation of systematically reusable software components,
  (b) increased interoperability among software components, and
  (c) convenient and ready adaptation of software components.

More than a decade later, many of the advantages NIST ATP identified have not been realized in spite of significant research investments. Component-based software development remains an "unsolved problem" largely due to the vast and varied landscape of programming languages, operating systems, and hardware available.

Component-based development of M&S applications may also be considered an "unsolved problem" due to several factors:

(1) Components that need to be assembled with each other are coded in different programming languages intended to run under different operating systems on different hardware platforms.
(2) The level of granularity and fidelity (degree of representativeness) provided in a component is not compatible when assembled with other components having different levels of granularity and fidelity.
(3) A component available only in binary form with no source code and documentation creates uncertainties when conducting verification and validation processes.
(4) The intended uses of a component do not match the intended uses of the other components when the components are assembled together.
(5) A component providing much more functionality than needed degrades execution efficiency.

The US Department of Defense has created a number of M&S repositories (APL 2010). Reusing, composing, or adapting resources from these repositories has been hindered because of (1) the differences in programming languages, operating systems, and hardware, (2) classified nature of many models and simulations and associated data, (3) lack of organizational push for reuse, (4) lack of contractors' interest in reuse, and (5) lack of effective documentation.

Within the software engineering community, reuse is considered by many to be a "solved problem" for *cloud-based software development* under the Service-Oriented Architecture (SOA). A software application can be implemented as a Web service, and other applications can reuse via XML or JSON communications. The programming language used in developing the software application, the operating system it runs under, and the server computer hardware it runs on are transparent to the calling application.

To effectively engender reusability, composability, and adaptability problems, the M&S community should pursue the Web services paradigms that have been successfully applied within the general software arena.

6.2.1.4 Quality-Centric Approaches to Component Evaluation

An existing M&S application can be reused without any change if and only if its credibility is substantiated to be sufficient for the intended reuse purpose.

An existing submodel (model component) can be reused without any change if and only if

(a) its credibility is substantiated to be sufficient for the intended uses for which it is created and
(b) its intended uses match the intended uses of the simulation model into which it will be integrated.

Any change to the M&S application will require it to be verified, validated, and certified again. Any change to the existing submodel will require not only the submodel, but also the entire simulation model to be verified, validated, and certified again.

Traditionally, verification and validation (V&V) are conducted to assess the *accuracy* of a model. However, *accuracy* is just one of dozens of quality indicators affecting the overall usefulness of an M&S application. Arguably, *accuracy* is the most important quality characteristic; however, we cannot ignore the importance of other quality indicators such as *adaptability, composability, extensibility, interoperability, maintainability, modifiability, openness, performance, reusability, scalability, and usability.*

It is crucially important that M&S application development be carried out under a *quality-centric approach* rather than just the traditional *accuracy-centric approach*. It should be noted that a quality-centric approach embodies the accuracy-centric approach since accuracy is a quality characteristic by itself.

New methodologies, approaches, and techniques must be created under a quality-centric paradigm for assessing the overall quality of an M&S application by employing quality indicators such as accuracy, adaptability, composability, extensibility, interoperability, maintainability, modifiability, openness, performance, reusability, scalability, and usability.

6.2.2   Data Reuse in Practice

The increasing volume, velocity, and variety of available data present both great opportunities and great challenges. This is true across all areas of government, academia, and industry and is equally true of models and simulations. The value of M&S can rely heavily on the availability and quality of input data. Similarly, M&S can be prolific sources of output data. The United Nations Economic Commission for Europe (UNECE) projects global data to reach 40 Zettabytes (40 Billion Terabytes) by 2019 (see Fig. 6.2). The business and culture of government at large, the defense industry in particular, and the modeling and simulation enterprise are suffering under the weight of this data "glut." Organizational practices for managing, analyzing, and sharing data are becoming increasingly ineffective in the face of the volumes of data they must contend with on a daily basis. New strategies, approaches, and technologies are needed to meet this challenge.

During a 2015 address, then President Obama noted that "Understanding and innovating with data has the potential to change the way we do almost anything for the better" (Strata + Hadoop World 2015). This begins by thinking about data differently. Data is the foundation of information, knowledge, and wisdom (see Fig. 6.3). Within the US Army, data are an enterprise asset, information is an enterprise currency, and knowledge is an enterprise resource (Office of the Army

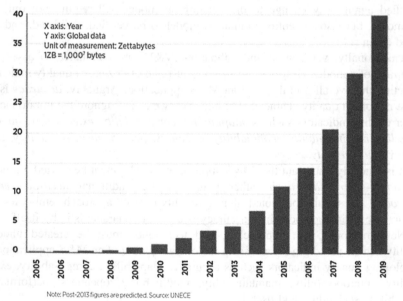

Fig. 6.2 United nations global data growth projections in Zetabytes

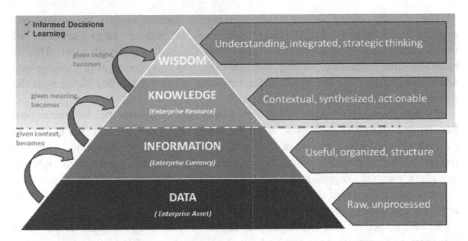

Fig. 6.3 DIKM pyramid levels of processing (Wikipedia, n.d.)

Chief Information Officer/G-6, February 2016). How we manage, analyze, and share data is what allows us to work our way up the information pyramid, such that we are using data and technology to "make a real difference in people's lives" (White House Office of Science and Technology Policy (OSTP), May 2012).

Today's amazing mix of cloud computing, ever-smarter mobile devices, and collaboration tools is changing the consumer landscape and bleeding into government as both an opportunity and a challenge. We describe research challenges in data reuse in three areas: management, analysis, and sharing.

6.2.2.1 Managing Data

Data management practices refer to the storing, identifying, organizing, correcting, and validating data. Vast amounts of data are produced across government, defense, and modeling and simulation (M&S) organizations. Currently, many models and simulations use a series of inconsistent ad hoc data structures to log and store data based on legacy files and formats. These data structures range from flat files to relational/hierarchical databases across a multitude of formats and from an even wider variety of sources. In addition, the data structures for each simulation event are unique and do not include metadata descriptions making direct comparison of data between events extremely difficult. Consider the challenge of maintaining and updating ad hoc data structures from simulation-supported event to simulation-supported event and how it would impact model and/or simulation scenario development, integration, analysis, verification, and validation of the models, simulation and data generated for a single system. Now, consider this challenge across a system of systems. It could quickly turn into a time-intensive effort that produces questionable results of limited value that impact the credibility of the models, simulation, and data produced.

Traditionally, after simulation-supported events, users of models and simulations employ a technique called *data reduction* to reduce large amounts of multi-dimensional data down to a corrected, ordered, simplified form. This is typically done by editing, scaling, summarizing, and other forms of processing into tabular summaries. In this process, the raw data is often discarded along with the hidden knowledge one may obtain from it.

As data storage costs have fallen, data management opportunities have emerged. Society has started to shift away from "minimal data storage" concepts to storing everything—including raw data that was once subjected to reduction. This has magnified the need and opportunity for more robust and advanced technologies in metadata creation, organization, and validation, necessary to produce properly formed data sets, sorted for processing, along with the data model required for the analysis and sharing aspects of data reuse.

Effective, active data management practices will promote data reuse, data integrity, and complex analytics and are the foundations of Data Science.

6.2.2.2 Analyzing Data

Data analysis refers to the process of inspecting data with the goal of discovering useful information, suggesting conclusions, and supporting decision making. In other words, how to most effectively move decision makers to the top of the knowledge pyramid where wisdom can be applied to accomplish goals by making decisions. *Exploratory, inferential*, and *predictive* data analytics are the three main bodies of analysis used in modeling and simulation.

In exploratory data analysis for simulations, the goal is to describe the data and interpret past results. These are generalizations about the data that are good for discovering new connections, defining new M&S scenarios for testing, and grouping observed events into classifications. Examples include capturing the number of successes versus number of attempts in a simulation, applications of census data, and number of times events occurred when particular conditions were met.

In inferential data analysis for simulations, the goal is to make inferences about the systems behavior based on a limited number of simulation events. This is critical due to the complexity and computational resources required to test every variable parameter for every possible condition within a complex system or system of system simulation. Examples include applications of polling and other sampling methodologies.

In predictive data analysis for simulations, the goal is to use previously collected data to predict the outcome of a new event. This helps with gaining deeper understandings of the interactions of the complex systems or system of systems. This involves measuring both the quantity and the uncertainty of the prediction. Examples include making predictions of future behavior based on credit scores, Internet search results, and so forth.

As our systems become more interconnected, and the models and simulations of these systems become more sophisticated, richer and richer data sets are produced. These systems are often loosely coupled, composed of multi-mission and multi-role entities, across organizations and often display non-obvious behaviors when operating in a complex environment. Exploring these relationships is a key component of a technique called data mining. Data mining complex simulations typically involves four common classes of tasks: anomaly detection, clustering, classification, and regression. Anomaly detection is focused on data errors or outliers that may be interesting to an analyst, e.g., failure modes. Clustering is a method of assigning similarity scores to groupings of like events. Classification is the task of generalizing known structures to apply to new data, for example, predicting causes and effects on system performance. Regression attempts to find a function, or simplified model, that can describe the behavior of the system with the least amount of error.

Research opportunities relating to reuse and data analysis include common data visualization methods, applications of synthetic data, and realtime data streaming and processing.

6.2.2.3 Sharing Data

Sharing applies at all levels of the pyramid depicted in Fig. 6.3. Sharing of data, information, knowledge, and wisdom across the pyramid has several different challenges. One of the biggest challenges, which may be immutable, is culture. For example, it has been argued that a culture of not sharing contributed significantly to the intelligence and homeland security failures associated with 9/11. Subsequent to that event, the US intelligence community was reformed, restructured, and given a mandate to share information in an effort to change the culture. A similar "sharing culture" is needed for M&S data. Many would posit that despite existing mandates to share M&S data, sharing does not consistently occur *within* government departments (even for data which has been carefully characterized for limitations and constraints), let alone consistently occur *across* the entire federal government. Such can be the challenges of policy versus implementation. For example, within the Department of Defense, sharing is mandated by DoD Directive 8320.02, yet rarely or consistently is M&S data published (i.e., made accessible) to anyone beyond the very focused end user. The unintended consequence of not sharing across the M&S domain is the reproduction of data and results which may already exist elsewhere that could have been made available and accessible for reuse. Further discussion on the cultural barriers to sharing appears in Sect. 6.3.

This illustration shows the work an organization should do on its data but does not show the sharing implications. One implication is that organizations are unaware of the fact that they do not have good visibility into their own data. They do not know what they do not know. As shown in Fig. 6.4a, a majority of an organization's data typically remains either unused (discovered but not used) or with unknown (not discovered and not used) value. Once this shortcoming is overcome,

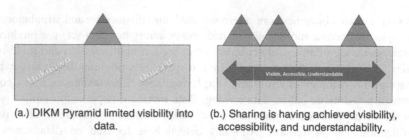

(a.) DIKM Pyramid limited visibility into   (b.) Sharing is having achieved visibility,
           data.   accessibility, and understandability.

Fig. 6.4 DIKM pyramid implications for sharing (Wikipedia, n.d.)

sharing can begin by achieving enterprise wide attributes of visibility, accessibility, and understandability (see Fig. 6.4b).

Conceptually, sharing of data defined as the data becoming visible, accessible, and understandable is quite simple and most organizations can tap the technology to achieve at least visibility and accessibility with commonly available information technology tools at their disposal. However, understandable data is more difficult to achieve. As mentioned previously, analysis is the gateway to understanding data beyond statistical summary, but it is not the solution for everything. The missing ability for universally understandable data is automatic semantic exchange across systems which access data. To share is to communicate. C.E. Shannon accurately identified the challenge (Shannon 1948):

> The fundamental problem of communication is that of reproducing at one point either exactly or approximately a message selected at another point. Frequently the messages have meaning; that is they refer to or are correlated according to some system with certain physical or conceptual entities. These semantic aspects of communication are irrelevant to the engineering problem.

In other words, the challenge of automated semantic data exchange (i.e., sharing of messages with meaning) not only depends on the data being shared but being properly interpreted contextually and properly used. The National Science Foundation is pursuing universal internet data models. Other approaches include work on automatic generation of taxonomies and reasoning mechanisms.

As sharing and collaboration gain traction within a community of interest, data security and privacy challenges naturally emerge. To understand the implications of data security, it is important to consider the states in which data exists. Data can be defined as being in motion (data transmitted over a connection), at rest (persistent for any length of time in any form), or in memory (in use by any program, tool, operating system, etc.). Threats to data in these forms can generally be classified into categories of privacy (unauthorized disclosure), integrity (alteration), and destruction (permanent deletion).

Although destruction is always a concern for data in any setting, of particular concern in a data sharing paradigm is privacy and integrity. The key emergent risk even in a "secure" sharing environment in a sharing paradigm becomes misuse. Data producers have an inherent concern that simulation data will not be used within the scope for which they pertain. Additionally, data producers may have a concern that if not kept private, its misuse could reflect poorly on the producer themselves (inaccurate embarrassment). Data integrity concerns arise for data producers with any reuse fearing alteration of original data. For these reasons in a data sharing paradigm; security is a key area of the present day and future research interest as the complexity extends far beyond general business use cases to highly technical scenarios. Consider, for example, that a transport security layer may be implemented to protect data in motion between two servers, as accessed by several different authenticated role-based users from different geographic locations, and that same data might also become vulnerable over the net by a distributed cache; secure in two ways (in motion and private) yet vulnerable in a third (in memory).

Data security and privacy need to be considered throughout the data life cycle in all of its mediums (as exampled above) and as technology around data is quickly changing such that security/privacy technology need to adjust to keep pace. Data security and privacy need to be considered as a part of risk management processes as well in such a way that they are not statically defined, but allow for continual evaluation for adapting new technologies ensuring they provide proper protections and safeguards to prevent improper collection, retention, use, or disclosure of data. Research is needed to support the creation of architectures where trust, account-ability, and transparency with respect to data collection, use, sharing, and security can be determined.

In conclusion, data is fundamentally changing how we conduct business and live our lives through advances in data quality, availability, and storage capacity. The M&S domain is no exception. Research opportunities abound in the Data Sciences, from metadata generation, organization, and validation within each data management category, to data analysis visualization, cloud computing impact on analysis, to data reuse in data analysis category, to automatic generation of taxonomies and reasoning mechanisms, effective security, and privacy in a sharing category just to name a few. Tremendous volumes of data enabled by technology for managing, analyzing, and sharing will drive new strategies and approaches that empower users to "connect the dots" through robust M&S and make well-informed decisions. We can now ask questions of the data that we never could before, even questions about the data itself such as should we throw data away and if so, when? The under-standing and motivation to embrace this reality across government, defense busi-ness, and culture needs to be applied to the M&S domain.

6.2.3 Discovery and Knowledge Management of M&S Assets

Knowledge management is a key area for M&S reuse with the recognition that models and simulations are encapsulations of knowledge. Although not generally considered in those terms, a model is a representation or interpretation of an object or physical phenomenon which engenders the understanding of the original. The applicability of models lies in a consistent and coherent representation of the knowledge about the thing or process being modeled.

Luban and Hincu (2009) emphasize the coupling between simulation and knowledge management. In referencing some earlier work, they state that "although the literature separates simulation and knowledge management, a more detailed analysis of these areas reveals that there are many links between them. More knowledge about the system can be discovered during simulation modeling process, and model development can be facilitated by collaborative knowledge management tools."

The following sections will consider some potential research areas related to knowledge management as an enabler of modeling and simulation.

6.2.3.1 Monitoring and Assessment of Running Simulations

The complexity of modern simulations makes it increasingly difficult for a human being to adequately understand the interconnections and dependencies to assure that the model or the simulation is a correct implementation. As models and simulations become adaptive, even persistent (e.g., a simulation that monitors current system state and serves as a real-time decision aid of a manufacturing process), assessing validity will require another application to monitor and analyze the running system. Such systems are often referred to as Dynamic Data-Driven Application Systems (DDDAS).

Research topics include non-intrusive monitoring; visualization of simulation process from multiple perspectives; and understanding interaction dependencies in distributed computing environments.

6.2.3.2 Model Specification, Construct Validity, Selection, and Credibility

Models are a substitute for the actual system which the researcher is interested in understanding. As such, the models have a degree of abstraction that requires some assessment of the quality of the model for the intended purpose. Balci (2004) describes quality assessment as "situation dependent since the desired set of characteristics changes from one M&S application to another. M&S application quality is typically assessed by considering the M&S application requirements, intended users, and project objectives."

The Capability Maturity Model Integration (CMMI) defines a process that is intended to assure that the resultant modeling and simulation application is well defined and documented. Of course, this assumes that the CMMI process was followed and executed by individuals that were trained and followed the proscribed procedures. Assigning a level of validity based on the model construct would generally require a test or demonstration that the model is consistent with the actual object or process. The specification and construction validity of a model or simulation is therefore highly dependent on the process and underlying assumptions used in the formulation of the model.

The selection of a model by a user is therefore based on the confidence of the user that the model was designed, constructed, and assessed by a set of reasonable criteria including a quality control process. Since there is generally no currently accepted numerically defined method for applying a confidence figure (e.g., 95%), it is left to the user to determine whether the model or simulation is sufficient for their purposes. The confidence that the user has in the model is also reflected by the validity associated with that model. Research topics include numerical confidence of quality and measures of appropriateness.

6.2.3.3   Attribute Labeling with Authoritative Vocabulary

In the area of reuse of models, the major stumbling block is generally language. A simple example would be building a model car. The instructions for assembly might include the phrase "attach the hood to the support on the firewall." This phrase is perfectly understandable to someone familiar with American English; however, the term "hood" in the UK would be replaced with "bonnet." The same piece serves the same function but utilizes different labels.

When someone constructs a model which does not use the language of mathematics, attribute labeling, descriptions, etc., become problematic. In addition, the modeler is influenced by their domain expertise which shapes a person's view of the world. Looking at a stream bed, a hydrologist considers the erosion factors due to grazing impacts, while the geologist considers the subsurface geology as a contributing factor.

Model reuse will be a consistent problem if there is no agreed upon way of describing a particular component of the model or the model itself. Research in this area focuses on the attribute labeling through the use of an authoritative or controlled vocabulary.

Patricia Harping (2010) with the Getty Research Institute provides a good example of the need for a controlled vocabulary for managing their art collections and the need for both descriptive and administrative data.

> Data elements record an identification of the type of object, creation information, dates of creation, place of origin and current location, subject matter, and physical description, as well as administrative information about provenance, history, acquisition, conservation, context related to other objects, and the published sources of this information. [...] Art and cultural heritage information provides unique challenges in display and retrieval. Information must be displayed to users in a way that allows expression of nuance, ambiguity, and uncertainty. The facts about cultural objects and their creators are not always known or straightforward, and it is misleading and contrary to the tenets of scholarship to fail to express this uncertainty. At the same time, efficient retrieval requires indexing according to consistent, well-defined rules and controlled terminology.

How does the above relate to models and simulations? Models, as with art, require descriptions of what they are and how they are intended to be used as well as what inputs and outputs are required. The use of an authoritative vocabulary that would allow both within-domain and cross-domain identification and usage would be exceedingly useful for model reuse. It must be recognized that a controlled vocabulary such as used in many database applications would be inadequate for model development. As stated above, the domain of the user influences choice in the modeling process. The vocabulary should therefore be derived from the model domain while providing consistency in the description or labeling. Research topics include the development of ontologies and authoritative vocabularies.

6.2.3.4 Context Management

Context as defined by the Merriam-Webster dictionary is the interrelated conditions in which something exists or occurs (e.g., environment, setting). For the development of models and simulations, there should exist a contextual framework which captures key model and simulation attributes. From a reuse standpoint, the contextual framework is essential for proper understanding of why the model was developed and how it was expected to be used.

A potential list of attributes that make up a contextual framework for a model might include the following:

1. Assumptions: What has been taken to be true?
2. Constraints: What conditions or restrictions have been applied to construction or use?
3. Intention: What was the purpose for which the model was developed?
4. Usage risk: What are the established and generally accepted application boundaries?
5. Model fidelity: What is the accuracy with which the model replicates the original?
6. Trust/Confidence: What measure can be used to assure the user of proper operation?
7. Pedigree/provenance: Who constructed or altered the model, what was changed, why was it changed, how was it changed, when was it changed?

The contextual framework provides the user community with enough information to make informed decisions, and this is a critical aspect in model reuse. In general, very little information is available with models, including ones which might be used to determine a life or death situation. It is an anecdotal assumption that many models in use today were developed by someone who is dead and no one knows how the model was constructed. Reconstruction of the contextual framework can be done to some extent, but will only result in a partial set of knowledge. Techniques for automatic context generation are needed.

6.2.3.5 Domain Knowledge Extension for Collaboration and Enhanced Decision Making

The primary use of models and simulation is to understand or convey information concerning a thing, process, or theory. The end product is a decision by the user as to the credibility of the model or simulation. The decision maker, whether an individual or group using models and simulations as a decision aid, must therefore believe that the results are reflective of the domain of interest. In cases where the model is obviously incorrect (e.g., water flows uphill), the results would be discarded by the decision maker. However, when the model or simulation provides what appears to be a correct result or at least a result which on the surface appears correct, how much credibility should be attributed to the model output?

Blatting et al. (2008) make the case that "... since credibility is subjective, different decision makers may well assign different degrees of credibility to the same M&S results; no one can be told by someone else how much confidence to place in something. The assessment of M&S credibility can be viewed as a two-part process. First, the M&S practitioner makes and conveys an assessment of the particular M&S results. Then, a decision maker infers the credibility of the M&S results presented to them in their particular decision scenario."

6.2.3.6 Reuse Through Model Discovery

Scudder et al. (March 23–27 2009) made the case that discovery would be achieved only through consistent and relevant metadata which in turn requires consistent labeling and a markup syntax. The basic concept is similar to other data discovery approaches which rely on the developer community to adhere to a defined set of rules for describing their models. A problem with standardizing model descriptions is getting agreement across modeling domains.

One definition of a model is representation of something (e.g., a system or entity) by describing it in a logical representation (e.g., mathematical, CAD, physical). The nature of the representation leads to the problem of discoverability as a form of the description. For example, how would you embed the necessary information in a mathematical model that would allow for discovery? The use of an indirect association (think of a catalog) would provide an access point but would also require a governance process that would maintain the associations as more models are added or if the models change.

Taylor et al. (2015) identified another aspect of reuse which is that some models would require specialized knowledge to use. As models have become more ubiquitous, the model interfaces have become easier to use. When the user was required to construct a very specific formatted file of input data, they had to understand the data and how it would be used in the model. With the graphic interface, it is easy to construct a model that anyone could run, but that does not insure that the output results would be valid. In addition, if the user discovered six models purporting to provide the same answers, how would the user determine which model is the best or most accurate?

The same authors make the case for reuse through standard ontologies and data models. The development of ontologies can be problematic due to the need for consensus among the domain subject matter experts. The initial barrier of developing the ontologies is outweighed by the significant gain in reuse and composability due to inherent relationship mapping.

6.3 Advancements in the Social, Behavioral, and Cultural Aspects of Reuse

Even when all the strictly technical challenges of reuse have been resolved, social, behavioral, and programmatic barriers may still prevent realization of the full potential of reuse. The social and behavior challenges identified in this subsection must be addressed with, if not ahead of, the technical challenges identified in the rest of this section of the report. There will be no substantive progress against the reuse challenge as a whole if the current workforce members who are most successful or influential within their respective domains are not concerned about M&S as a larger discipline. All of which is unlikely to happen unless funding comes to pay people to work the larger, broader, philosophical, and theoretical understanding of models, what they are, and how they work.

This subsection addresses identifying and teaching the skills necessary for a model or simulation producer to increase the ease of reuse by others if the producer (person or organization) chooses to and can afford to do so. So long as designing and documenting for reuse are not required and funded, these actions will be difficult to justify in the current contractual M&S culture.

In comparison with research on technical challenges, research in this area will require human experimentation, e.g., design of educational materials and testing of efficacy. Some challenges in this area may benefit from research into mental models of software users and developers in other communities including the open source community.

6.3.1 Programmatic

The observations in this section are largely based on experiences within the US DoD M&S community.

6.3.1.1 Governance

The US federal government and especially the Department of Defense (DoD) are significant consumers of M&S and commensurately stand to benefit the most from increased reuse. Federal procurement and acquisition policy currently reward non-reuse behavior and, in some ways, punishes[1] design and implementation for reuse. While the resolution of these issues is outside the scope of technical research, recognition of them is key when considering mechanisms for improving social and behavioral aspects, especially for identifying the strict limits policy imposes on the efficacy of activities to shape behavior change. The CNA report (Shea and Graham 2009) is a good source for understanding this issue. This barrier may not exist or be considerably lower outside the federal government market place.

6.3.1.2 Return on Investment (ROI)/Cost Benefit Analysis

Decision makers often ask about a new endeavor, "What's the ROI?" Answering this question in a manufacturing context where there are clear metrics of increased cost for process changes and (presumably) reduced unit costs is straightforward. The answer is considerably less clear in a context of cost avoidance, where the precise cost of producing a model or simulation without reuse might be unknowable. The research challenges in this area include the following:

1. Defining a broadly acceptable framework for performing cost benefit analysis for reuse that does not rely on unknowable metrics.
2. Recognizing that different domains may have different mechanisms of practice, i.e., different protocols for reuse. These differences might lead to differing measures of effectiveness and/or methods for assessing effectiveness.

6.3.2 Risk and Liability

The risks and liabilities of reusing M&S developed by other organizations and/or for other domains and intended uses are too numerous to cite here. The challenge in this area is to determine where existing legal precedent applies and where new law must be established, an enterprise that can only be undertaken in collaboration with legal professionals, not by technical experts alone. Technical experts may contribute to this undertaking by providing and developing appropriate mechanisms for assessing the technical aspects of risk and failure. A notable example of extant work in this area is the Risk-Based Methodology for VV&A (The Johns Hopkins Applied

---

[1]Title 31 US Code § 1301 restricts the use of current funds to fund future anticipated, but not yet realized, requirements.

Physics Laboratory, April 2011). Intellectual property (IP) rights are also a consideration in this area. A user may discover the need to modify a reusable asset for their specific intended use. Without acquiring appropriate IP rights prior to reusing the asset, the user will have (unintentionally) accepted a risk or liability that is costly to mitigate. The CNA report (Shea and Graham 2009) covers this topic in some detail.

6.3.3 Social and Behavioral

6.3.3.1 Motivating Behavior Change

Recognizing the constraints imposed by governance, reuse can only succeed through shaping changes in stakeholder behavior and decision making regarding reuse.

Specific research challenges associated with the social behaviors of producers, consumers, integrators, and decision makers necessary to build and sustain a viable community of reuse include the following:

- *Design for reuse.* What reward structures and/or response costs encourage this behavior? Could approaches such as gamification create a positive feedback loop between individual and group behavior? What skills and/or techniques are necessary to achieve reusable designs? What infrastructure and mechanisms are necessary to provide constructive feedback to designers of reusable assets?
- *Documentation for reuse.* Even when an asset is designed for reuse, failure to provide sufficient documentation, especially discovery and composition metadata, limits its reusability. It is not uncommon for software developers to resist producing sufficiently detailed and informative documentation once the code is working. The lack of documentation impedes verification and validation (V&V), and subsequently, reuse. The challenges in this area are similar to those in design for reuse, but require separate consideration.
- *Reusing/adoption of reusable resources.* What reward structures/incentives encourage this behavior in the absence of governance constraints?
- *Implied threat of reuse to potential stakeholders.* Reward structures and incentives represent the positive side of encouraging reuse, but reuse can also represent an implied threat to potential stakeholders, e.g., loss of funding, control, and/or perceived status. Research in this area needs to identify implied threats, and assess whether rewards and incentives can counteract them. While trust may not directly counteract perceived threats, it may ameliorate them. In this context, trust applies to individuals, organizations, accuracy of metadata, and quality of reusable assets.
- *Different levels/types of motivation and concerns.* Research in the preceding areas must account for the fact that different stakeholders will have different levels and types of motivations and concerns.

6.3.3.2 Education and Outreach

Finally, the results of the research described in the preceding subsection must be delivered to the target audience(s) and measured for efficacy. This research should address the varying challenges of identifying target students, delivering effective education, and measuring its efficacy based on the students' roles within a community of reuse as: producers, consumers, integrators, or decision/policy makers.

The education and evaluation process should investigate various outreach mechanisms including expert endorsements, and social media and networking. The LVCAR Asset Reuse report (APL 2010) describes several such mechanisms.

6.3.4 Impact

Addressing the research challenges identified in this subsection has the potential to achieve the following positive impacts:

- Creating a verifiable body of knowledge and standardized processes for calculating the benefits of reuse.
- Motivating a culture of reuse and rewarding stakeholders who engage constructively.
- Providing stakeholders with constructive methods for overcoming resistance.

The cultural norms resistant to reuse are entrenched and unlikely to change without arming individuals motivated to change it with concrete tools.

References

ACM SIGSIM. 2016. *Modeling and Simulation Knowledge Repository (MSKRR).* http://www.acm-sigsim-mskr.org/MSAreas/msAreas.htm. Accessed 20 Apr 2016.

APL. 2010. Live-Virtual-Constructive Common Capabilities: Asset Reuse Mechanisms Implementation Plan. Laurel, MD. Applied Physics Laboratory, The Johns Hopkins University, Technical Report. NSAD-R-2010-023.

Balci, O. 1998. Verification, validation and testing. In *Handbook of Simulation: Principles, Methodology, Advances, Applications and Practice.* New York, NY: John Wiley & Sons, 335–393.

Balci, O. 2004. Quality assessment, verification, and validation of modeling and simulation applications. In *Proceedings of the Winter Simulation Conference.*

Balci, O. 2016. Introduction to modeling and simulation. ACM SIGSIM Modeling and Simulation Knowledge Repository (MSKRR) Courseware. Accessed 20 April 2016. http://www.acm-sigsim-mskr.org/Courseware/Balci/introToMS.htm.

Balci, O., J.D. Arthur, and W.F. Ormsby. 2011. Achieving reusability and composability with a simulation conceptual model. *Journal of Simulation* 5 (3): 157–165.

Blatting, S.R., L.L. Green, J.M. Luckring, J.H. Morrison, R.K. Tripathi, and T.A. Zang. 2008. Towards a capability assessment of models and simulations. In *49th AIAA/ASME/ASCE/AHS/ASC Structures, Structural Dynamics, and Materials Conference.*

Clark, J., C. Clarke, S. De Panfilis, G. Granatella, P. Predonzani, A. Sillitti, G. Succi, and T. Vernazza. 2004. Selecting components in large COTS repositories. *The Journal of Systems and Software* 73 (2): 323–331.

Fujimoto, R.M. 2016. Research challenges in parallel and distributed simulation. *ACM Transactions on Modeling and Computer Simulation.* 24(4) (March 2016).

Harkrider, S.M., and W.H Lunceford. 1999. Modeling and simulation composability. In *Proceedings of the 1999 Interservice/Industry Training, Simulation and Education Conference.* Orlando, FL.

Harping, P. 2010. Introduction to controlled vocabularies: terminology for art, architecture, and other cultural works. Murtha Baca, Series Editor, J. Paul Getty Trust.

IEEE. IEEE Standard 1516, 1516-1, 1516-2, and 1516-3. IEEE Standard for Modeling and Simulation (M&S) High Level Architecture (HLA). New York, NY.

Luban, F. and D. Hincu. 2009. Interdependency between simulation model development and knowledge management. *Theoretical and Empirical Researches in Urban Management,* 1(10).

Mahmood, I. 2013. A verification framework for component based modeling and simulation: putting the pieces together. Ph.D. Thesis, KTH School of Information and Communication Technology.

Modeling and Simulation Coordination Office (MSCO). 2016. DoD M&S Catalog. Accessed 21 Apr 2016. http://mscatalog.msco.mil/.

NIST. 2005. *Component-Based Software.* Washington, DC: Advanced Technology Program (ATP) focused program.

Office of the Army Chief Information Officer/G-6. 2016. *Army Data Strategy.* Washington, DC.

Page, E. H., and J. M Opper. 1999. Observations on the complexity of composable simulation. In *Proceedings of the 1999 Winter Simulation Conference.* Phoenix, AZ.

Page, E.H., Briggs, R., and Tufarolo, J.A. 2003. Toward a family of maturity models for the simulation interconnection problem. In *Proceedings of the Spring 2004 Simulation Interoperability Workshop.* Arlington, VA.

Petty, M.D. 2004. Simple composition suffices to assemble any composite model. In *Proceedings of the Spring 2004 Simulation Interoperability Workshop.* Arlington, VA.

Petty, M.D. and Weisel, E.W. 2003. A composability lexicon. In *Proceedings of the 2003 Spring Simulation Interoperability Workshop.* Orlando, FL.

Petty, M.D., E.W. Weisel, and R.R Mielke. 2003a. A formal approach to composability. In *Proceedings of the 2003 Interservice/Industry/Training, Simulation and Education Conference.* Orlando, FL.

Petty, M.D., E.W. Weisel, and R.R Mielke. 2003b. Computational complexity of selecting models for composition. In *Proceedings of the Fall 2003 Simulation Interoperability Workshop.* Orlando, FL.

Petty, M.D., J. Kim, S.E. Barbosa, and J. Pyun. 2014. Software frameworks for model composition. *Modelling and Simulation in Engineering.*

Petty, M.D., K.L. Morse, W.C. Riggs, P. Gustavson, and H. Rutherford. 2010. A reuse lexicon: terms, units, and modes in M&S asset reuse. In *Proceedings of the Fall 2010 Simulation Interoperability Workshop.* Orlando, FL.

Pressman, R.S., and B.R. Maxim. 2015. *Software Engineering: A Practitioner's Approach.* New York, NY: McGraw Hill.

Sabah, M., and O. Balci. 2005. Web-based random variate generation for stochastic simulations. *International Journal of Simulation and Process Modelling* 1 (1–2): 16–25.

Scudder, R., P. Gustavson, R. Daehler-Wilking, and C. Blais. 2009. *Discovery and Reuse of Modeling and Simulation Assets.* Orlando, FL: Spring Simulation Interoperability Workshop.

Shannon, C.E. 1948. A mathematical theory of communication. *Bell System Technical Journal* 27 (3): 379–423.

Shea, D.P. and K.L. Graham. 2009. Business models to advance the reuse of modeling and simulation resources. Washington, DC: CRM D0019387.A2/Final, CNA.

Strata + Hadoop World 2015. Accessed 29 April 2016. https://www.youtube.com/watch?v=qSQPgADT6bc.

Taylor, S. et al. 2015. Grand challenges for modeling and simulation: simulation everywhere–from cyberinfrastructure to clouds to citizens. *Transactions of the Society for Modeling and Simulation International*, 1–18.

The Johns Hopkins Applied Physics Laboratory. 2011. Risk Based Methodology for Verification, Validation, and Accreditation (VV&A). M&S Use Risk Methodology (MURM). Washington, DC: NSAD-R-2011-011.

Tolk, A. and J.A. Muguira. 2003. The levels of conceptual interoperability model. In *Proceedings of the Fall 2003 Simulation Interoperability Workshop*. Orlando, FL.

Tolk, A., S.Y. Diallo, J.J. Padilla, and H. Herencia-Zapana. 2013. Reference modelling in support of M&S — foundations and applications. *Journal of Simulation* 7 (2): 69–82.

Weisel, E.W., R.R. Mielke, and M.D. Petty. 2003. Validity of models and classes of models in semantic composability. In *Proceedings of the Fall 2003 Simulation Interoperability Workshop*. Orlando, FL.

White House Office of Science and Technology Policy (OSTP). 2012. Digital government: building a 21st century platform to better serve the American people. Washington, DC.

Yilmaz, L. and T.L. Oren. 2004. A conceptual model for reusable simulations with a model-simulator-context framework. In *Proceedings of the Conference on Conceptual Modelling and Simulation*. Genoa, Italy.

Erratum to: Computational Challenges in Modeling and Simulation

Christopher Carothers, Alois Ferscha, Richard Fujimoto,
David Jefferson, Margaret Loper, Madhav Marathe,
Pieter Mosterman, Simon J.E. Taylor and Hamid Vakilzadian

Erratum to:
Chapter 4 in: R. Fujimoto et al. (eds.), *Research Challenges in Modeling and Simulation for Engineering Complex Systems*, Simulation Foundations, Methods and Applications, DOI 10.1007/978-3-319-58544-4_4

In the original version of the book, the contributor name and the corresponding affiliation have to be included in Chapter 4. The erratum chapter and the book have been updated with the change.

The updated original online version of this chapter can be found at
https://doi.org/10.1007/978-3-319-58544-4_4

© Springer International Publishing AG (outside the USA) 2017   E1
R. Fujimoto et al. (eds.), *Research Challenges in Modeling
and Simulation for Engineering Complex Systems*, Simulation Foundations,
Methods and Applications, DOI 10.1007/978-3-319-58544-4_7

Index

A
Acquisitions, 14–16, 25, 34
Adaptation, 10, 59, 97, 98
Aggregation of information, 81
Aggregation of preference, 81
Array processor*See*FPGA, 52, 53

B
Bayesian probability theory, 76, 77
Bias in judgement of uncertainty, 7
Big data, 2, 47, 52, 66, 84

C
Capability Maturity Model Integration
    (CMMI), 34, 35, 106
Climate change, 14, 15
Cloud computing, 2, 47, 53, 54, 84, 101, 105
Communicating uncertainty, 83
Communication among multiple stakeholders,
    24
Component, 6, 15, 16, 19, 20, 35, 39, 40, 50,
    54, 58–62, 64, 65, 68, 88–94, 96–99,
    103
Composability, 38, 89–91, 93, 94, 97, 99, 110
Composition, 10, 18, 20, 39, 41, 81, 89–92, 94,
    96, 97, 112
Computational complexity, 28, 89
Compute Unified Device Architecture
    (CUDA), 53
Conceptual model, 3, 5, 8, 10, 19, 24–32,
    35–40, 96
Context, 3, 6–10, 14, 17, 20, 24, 36, 37, 47,
    57–59, 67–69, 71, 76, 78, 80, 84, 88, 92,
    107, 108, 111, 112

Co-simulation, 41, 64, 65
Credibility, 17, 29, 32, 35, 99, 101, 106, 109
Cyber-physical systems, 47, 56, 59, 64, 65

D
Data analysis, 47, 52, 84, 102, 103, 105
Data science, 102, 105
Decision making, 15–17, 58
Decision theory, 37, 78
Discovery, 88, 93, 105, 109
Discrete event simulation, 6, 28, 48, 49, 51, 64
DoDAF, 26, 35
Domain-specific Modeling Languages
    (DSMLs), 26, 27
Dynamical systems, 28
Dynamic Data Driven Application Systems
    (DDDAS), 6, 47, 57

E
Early stage models, 24, 27
Education of Uncertainty, 83
Emergent behavior, 17, 18
Energy consumption (by simulations), 2, 49,
    55, 56
ENIAC, 1
Execution integration, 41

F
Federated simulation models, 35, 61, 62
Federation Development and Execution
    Process (FEDEP), 35
Field Programmable Gate Array (FPGA), 47,
    48, 52
Formal modeling languages, 8

© Springer International Publishing AG (outside the USA) 2017
R. Fujimoto et al. (eds.), *Research Challenges in Modeling
and Simulation for Engineering Complex Systems*, Simulation Foundations,
Methods and Applications, DOI 10.1007/978-3-319-58544-4

Printed in the United States
By Bookmasters